PASSENGER AIRCRAFT and their INTERIORS 1910~2006

JOHN STROUD

SCOVAL
PUBLISHING LTD

Written By John Stroud

British Library Cataloguing in Publication Data
A catalogue record for this book is available from the British Library

ISBN: 1 902236 03 3

Published by:
SCOVAL Publishing Ltd
PO BOX 36
Ponteland
Newcastle-upon-Tyne
NE20 9WE
England
Tel: (01661) 820 838
Fax: (01661) 822 911

Printed by Kyodo Printing Co (S'Pore) Pte Ltd
Singapore

Edited by S Henderson and C Hymers
Produced and typeset in 11pt on 14pt Quorum
Designed by Scott Henderson, for SCOVAL Publishing Ltd

CONTENTS

INTRODUCTION

The history of aeronautical development has been extensively recorded including numerous works on the transport aeroplane and passenger-carrying airship (there are nearly 4,000 books on air transport in English alone). These works have described structures, engines and performance as well as in-service histories, but scant attention has, in general, been given to the essential passenger accommodation.

As we approach the centenary of air travel, I believe this gap in recorded aeronautical history should be closed.

There are a number of books on the development of the cockpit up to the modern airliner flight deck (the glass cockpit) so in this work I have concentrated on the passenger accommodation.

This book traces the development of the passenger aircraft interior from the first passenger aircraft, the Zeppelin LZ 7 Deutschland of 1910, and the first passenger-carrying aeroplanes through the 1920s and 1930s to the post-war piston-engined airliners and then the jet age up to the present with a glimpse into the future with an aeroplane designed to carry nearly a thousand passengers.

An explanation is necessary of apparent inconsistencies. For example Lufthansa was two words until the end of 1933, United Airlines was originally three words and Pan American Airways added World to its title after the Second World War. Fokker aircraft type designations had a full point after the F for aircraft built in Europe but a dash for those built in the United States.

ACKNOWLEDGEMENTS

In writing this work I have received much generous assistance and I offer my sincere thanks to Airbus Industrie; Alitalia; Dr Norman Barfield; Boeing; R E G Davies, Curator of Air Transport, National Air and Space Museum, Washington DC; Jim Ferguson; Harold Forrester; Scott Henderson; Mike Hooks; Lufthansa; Sir Peter Masefield; David Morgan; Ray Piercey; D M (Mike) Stroud; Swissair; the late John W R Taylor; Verkehrshaus der Schweiz and Angela Young.

John Stroud
Nairn, Scotland
May 2002

For Angela
and in
Loving Memory of My Patricia

CHAPTER ONE
1910 - 1914

The first passenger aircraft was the Delag Zeppelin LZ 7 Deutschland which first flew on 19 June 1910. Deutschland had seating for twenty passengers, was powered by three 120hp engines, 485ft 6in long, had a gas capacity of 681,574cu ft and a top speed of 37mph. (Luftschiffbau Zeppelin)

Delag Zeppelins

The first passenger-carrying aircraft were a number of rigid airships built at Friedrichshafen by Luftschiffbau Zeppelin Gmbh for its subsidiary Deutsche Luftschiffahrts AG (Delag) which was founded at Frankfurt-am-Main on 16 November 1909 to operate passenger flights and train Zeppelin crews.

Delag was thus the first airline although it did not operate scheduled services until after the 1914-18 war. Delag's first Zeppelin was the Type e LZ 7 Deutschland which first flew on 19 June 1910 and went into service on 28 June but was written off on its first voyage due to bad weather and poor airmanship. Deutschland was 148m (485ft 6in) long, powered by three 120hp Daimler engines, had 19,300cu m (681,574cu ft) gas capacity and a maximum speed of 59.5km/h (37.35mph). It had beneath its hull a tram-like cabin for twenty passengers.

After its accident Deutschland was replaced by the almost identical LZ 8 Ersatz Deutschland which first flew on 30 March 1911. It was soon destroyed when being walked out of its shed.

The two Type e Zeppelins were replaced by modifying the Type d LZ 6 which had been built in expectation of an Army order and first flew in August 1909. Modified as the LZ 6A it had accommodation for ten to twelve passengers and it began daily passenger flights in August 1910. It was destroyed by fire in its hangar that September.

On 15 July 1911 Delag took delivery of the Type f 'ship LZ 10 Schwaben. It had first flown on 26 June that year, was more successful than previous Zeppelins and had more reliable 145hp Maybach engines. Schwaben was the first aircraft to have a steward in its crew and it is known that he served cold meals and a selection of wines and champagnes. Before being wrecked by a storm at Düsseldorf at the end of June 1912 Schwaben had carried 1,553 revenue passengers. Two Type g Zeppelins went into service in 1912. They were LZ 11 Viktoria Luise and LZ 13 Hansa and were approximately the same size as the earlier 'ships' but with accommodation for 25 passengers.

Similar but having greater gas capacity, was the Type h LZ 17 Sachsen, the last of the pre-1914 Delag airships. Sachsen began regular operation in June 1913 and in Delag service was reported to have flown 740hr and

The passenger gondola of the Deutschland with windows open. (Luftschiffbau Zeppelin)

carried 9,837 passengers. Delag published maps of a network of German domestic routes but never operated pre-war scheduled services. There were some inter-city flights and Sachsen once flew to Vienna but most operations were cruises from the various airship harbours which the company had set up.

However, the Delag Zeppelins had a high standard of passenger accommodation. Walls and ceilings of the cabins were covered by mahogany plywood, the five-ply floors were covered by carpets and the wicker seats were arranged along the sides of the cabin with single facing seats in pairs either side of tables which had white table cloths. The large cabin windows could be opened and most, if not all, had wash rooms and lavatories. It was many years before accommodation in aeroplanes even approached the Zeppelin standards.

When the approaching war brought Delag operations to an end on 31 July 1914, the airline's Zeppelins had flown 172,535km (107,205miles) with 33,722 passengers and crew in 1,588 flights and over 3,000hr and no passenger was ever injured.

Le Grand

Although it never entered passenger service Igor Sikorsky's S-21 Bolshoi Baltiskiy (Great Baltic) must be mentioned. This 26m (88ft) span biplane was built by the Russian Baltic Railcar factory in St Petersburg during the winter of 1912-13.

Initially the S-21 had two 100hp Argus engines and in this form it flew on 2 March 1913. Two more Argus engines were added and it made the first flight by a four-engined aeroplane on 13 May 1913. Generally known as Le Grand (The Great One) this large aeroplane weighed 3,400kg (7,500lb) and had a full-height enclosed cabin with dual controls. The cabin aft of the pilots had wicker seats, tables, electric light, large windows, a coat-room and a lavatory. Le Grand is known to have flown with eight on board. Its weight was increased to 4,195kg (9,250lb) and it was renamed Russkiy Vityaz (Russian Knight). Unfortunately it was destroyed in August 1913 when an engine fell off an overflying aircraft and hit the port wings of the big Sikorsky.

The S-21 was followed by the S-22 Ilya Mourometz which had a lounge with wicker seats, tables and a lavatory. It was put into production but as a bomber.

Benoist flying-boat

On 1 January 1914 St Petersburg - Tampa Airboat Line operated the first-ever scheduled air service for passengers. The service connected St Petersburg with Tampa, both in Florida, a distance of 21 miles across Tampa Bay. The first flight took 23 minutes was piloted by Tony Jannus and carried one passenger.

The aeroplane used was a small wood and fabric Benoist biplane flying-boat which had a 75hp Roberts engine. The service operated each way twice a day and operation continued until the end of that March by which time more than 1,200 passengers had been carried.

The original Benoist passenger accommodation was a seat for one passenger beside the pilot in an open cockpit. After a while a second two-passenger Benoist was used. This Benoist service and the Delag Zeppelin flights were the only passenger carrying activities before the 1914-18 war.

Igor Sikorsky's Bolshoi Baltiskiy (Great Baltic) or Le Grand (The Great One) on 13 May 1913 made the first flight by a four-engined aeroplane. It had a full-height cabin and dual controls.

This model of the Benoist flying-boat clearly shows the open cockpit which had two seats, for the pilot and one passenger. (Qantas)

CHAPTER TWO
BODENSEE AND NORDSTERN

St Petersburg - Tampa Airboat Line's Benoist flying-boat making the world's first scheduled passenger flight on 1 January 1914, from St. Petersburg to Tampa, both in Florida. (Courtesy John WR Taylor)

Soon after the Armistice in November 1918, Delag began to plan for the resumption of airship operations and ordered a new Zeppelin to operate German domestic services.

Bodensee

The new airship was the Type y LZ 120 Bodensee (Lake Constance). It was designed in about three months and first flew on 20 August 1919. Bodensee differed in many ways from previous Zeppelins. The earlier Zeppelins had long constant-diameter hulls but Bodensee had a fully-streamlined hull, simplified but more effective control surfaces and the passenger cabin and crew accommodation and control area were in one under-hull structure instead of being separate units. Bodensee had four 260hp Maybach engines whereas earlier Delag 'ships had been three-engined. The new airship was shorter than the earlier passenger Zeppelins but had greater gas capacity than all except Sachsen. It was also faster, having a cruising speed of 120.7km/h (75mph) and a top speed of 133.2km/h (82.7mph).

The passenger cabin was much improved with twenty seats, arranged ten each side beside large windows with an extra seat at the rear of the control cabin. When the 'ship had sufficient lift, up to six wicker chairs were placed in the aisle bringing total seating to 27.

Delag had planned to run regular services between Friedrichshafen and Berlin with later extensions to Italy, Spain, Sweden and Switzerland. The Friedrichshafen-Berlin service began on 24 August 1919 with northbound and southbound flights on alternate days but international services were never worked. Initially the flights were via Munich and took about six hours but from early October the flights were non-stop and took about four hours. The last service was operated on 5 December by which time Bodensee had flown 51,258km (31,850miles) including a flight to Stockholm and carried 4,050 passengers of which 2,253 were fare paying. Bodensee was laid up for the winter and was lengthened by 10m (32ft 9.5in) in readiness for the 1920 season.

Nordstern

Delag had ordered a second Zeppelin to be used on a Berlin-Stockholm service. This was LZ 121 Nordstern (North Star) which first flew on 8 June 1921. Nordstern was almost identical to the lengthened Bodensee but had improved accommodation. The passenger cabin was in

The Delag Zeppelin LZ 120 Bodensee at Friedrichshafen in 1919. (Luftschiffbau Zeppelin)

The well appointed passenger accommodation in Nordstern. (LZ 121)

Part of the cabin of Bodensee with a Friedrichshafen—Berlin route map on the bulkhead.

two sections divided by semi-bulkheads. In each section there were on one side two groups of double seats facing across a table and single facing seats with tables on the opposite side. There were large opening windows, luggage racks and electric ceiling lights. Total seating was for 24 and the standard of passenger accommodation far surpassed that provided in any aeroplane for many years. However, Delag was not to have a 1920 season and Nordstern's first flight was delayed because the Inter-Allied Commission of Control told the German Government that the airships must not be flown but handed to the Allies as war reparations. Bodensee went to the Italian Navy as Esperiá and Nordstern became the French Navy's Méditerranée — they survived respectively until 1927 and 1926.

CHAPTER THREE
MILITARY CONVERSIONS

The Airco D.H.4A F5764 was one of those operated by No 2 (Communication) Squadron of the Royal Air Force

United Kingdom

After the Armistice, the Peace Conference was held at Versailles and rapid transport was required between London and Paris for members of the British Government and others. Royal Air Force Communication Squadrons were formed to undertake this work and they operated the flights with an assortment of Airco D.H.4s and D.H.4As and some Handley Page O/400s. Regular London-Paris military-operated services began on 10 January 1919.

Earlier, on 5 October 1916, George Holt Thomas, founder of The Aircraft Manufacturing Co. (Airco), had formed Aircraft Transport & Travel, the first British airline, with the object of 'establishing and maintaining lines' and to enter into contracts for the carriage of mails passengers, goods and cattle. However, the war was to continue for another two years and civil flying in the British Isles was only allowed from 1 May 1919. On that day, an Aircraft Transport & Travel D.H.9 made the first flight under the new regulations - from London with newspapers

The cabin area of a Communication Squadron D.H.4A. The access ladder is canted-out to clear the control wires. Entry and exit could not have been easy.

A D.H.16, probably the first with Napier Lion engine, at the 1919—20 Exposition Internationale de Locomotion Aérienne in the Grand Palais, Paris. The sign-board has the only known example of the name Airco being shown with a hyphen.

Passengers' baggage goes aboard an Aircraft Transport & Travel Airco D.H.16 at Hounslow in 1919 or 1920.

destined for Bournemouth, but the aircraft force-landed on Portsdown Hill because of fog.

Civil flying, in connection with the Peace Conference, was allowed between London and Paris from 13 to 20 July 1919 but regular operation was not allowed until August of that year.

On 25 August, A T & T inaugurated regular scheduled cross-Channel services from Hounslow, close to the present Heathrow, to Le Bourget, the Paris airport. On that day Handley Page Transport made a London-Paris flight but the airline did not begin regular operation until 2 September.

Passengers dressed for their journey in the open cockpit of a DDL D.H.9.

Airco D.H.4A

The Airco D.H.4 was designed by Geoffrey de Havilland and built in large numbers by The Aircraft Manufacturing Company and a number of sub-contractors. It was a wood and fabric single-engined two-seat biplane bomber which first flew in August 1916. D.H.4s were used to operate daily courier and mail services between England and France during the Peace Conference and the British Prime Minister, Mr Bonar Law, requested some D.H.4s be given enclosed accommodation for the carriage of Ministers and others and their secretaries.

Passengers about to board an Aircraft Transport & Travel Airco D.H.9 which was operating a service for KLM before the airline had its own aircraft.

This DDL — Danish Air Lines Airco D.H.9 has four open cockpits — three of them for passengers.

So instead of the open gunner's cockpit aft of the pilot space was made for two facing seats and a hinged cover was made to fold up from the starboard side. The cabin had two large windows in each side and was neatly faired into the rear fuselage. Entry and exit were made via a ladder attached to the starboard side between the two windows.

The D.H.4A's cabin was 6ft 3in long, had a depth of 4ft 8in and the approximate width was 2ft 7in.

Because of the backward shift of the centre of gravity, the rigging of the D.H.4's wings had to be modified and so the type was redesignated D.H.4A. At least seven D.H.4s were converted for the RAF Communication Squadrons and other examples were converted on the production line. Aircraft Transport & Travel had four and the Instone Air Line, as S. Instone and Co., began public service with a D.H.4A in February 1920. Handley Page Transport used one and SNETA (predecessor of Sabena) had two. One RAF and one A T & T

Passengers at Cricklewood wait to board Handley Page O/400s of Handley Page Transport. The open-nose cockpit had two passenger seats.

Handley Page O/400 G-EAAF after rebuild as an O/7.

Part of the interior of the Handley Page O/400 D8326 HMA Silver Star. This example and one or two other O/400s were bombers converted for VIP passengers attending the Peace Treaty talks at Versailles.

Looking aft in the cabin of a Handley Page O/7. There is a mixture of forward-and backward-facing wicker seats and fuselage structural members intrude into the cabin.

aircraft were lost in the English Channel and the Belgian aircraft were destroyed in a hangar fire at Brussels.

It has frequently been reported that it was the D.H.4A G-EAJC, flown by Lieut E H 'Bill' Lawford that inaugurated A T & T's cross-Channel services on 25 August 1919, but this is not true. Bill Lawford certainly made A T & T's first cross-Channel flight that day, leaving Hounslow at 09.05 but that was not the scheduled service that departed at 12.40, 12.30 being the actual scheduled time for the service.

Airco D.H.9

The D.H.9, also a single-engined two-seat biplane, was produced to carry heavier loads than the D.H.4 and to have increased range. It first flew in July 1917.

A T & T had several D.H.9s with open cockpits and DDL — Danish Air Lines, had at least one with four open cockpits - one for the pilot and three for passengers.

Numerous examples were fitted with enclosed cabins similar to the D.H.4A and some of these were designated D.H.9C. Some of KLM's first services were operated by open-cockpit D.H.9s of A T & T before KLM received any of its own aircraft.

Airco D.H.16

The Airco D.H.16 was the company's first type designed as a commercial aeroplane. The work's general arrangement drawing described it as 'D.H.16 Aerial Limousine seating pilot and four passengers'. Unfortunately there is no date on that drawing but the first of the nine examples built flew for the first time in March 1919. It and the next five were powered by 322hp Rolls-Royce Eagle VII water-cooled engines.

D.H.16s were built from D.H.9A components but given wider rear fuselages. The cabin had two forward- and two aft-facing seats arranged in semi-cockpits and the hinged cabin roof had two large windows each side plus fore and aft fixed windows. The work's drawing shows curtains and some aircraft may have had them. There were also windows in the roof. The cabin was 6ft 11in in length and had a width of 3ft 3in and access was via a ladder as on the D.H.4A. The last three had 450hp Napier Lion engines and one was experimentally fitted with

One of the best-known airliners of the 1920s was Instone Air Line's Vickers Vimy Commercial 'City of London', seen here taking-off from Croydon.

air-brakes and flaps. Eight of the D.H.16s were delivered to Aircraft Transport & Travel and it was the first of them, K-130 (later G-EACT), that flew the first scheduled cross-Channel service on 25 August 1919. Major Cyril Pattteson was the pilot and there were four passengers.

The third D.H.16, G-EALU Arras, operated the first London-Amsterdam service (joint A T & T and KLM) on 17 May 1920 with Lieut H 'Jerry' Shaw as pilot. One D.H.16 was sold to the River Plate Aviation Co in Buenos Aires and it operated services across the river to Montevideo in Uruguay. Two A T & T aircraft went to the de Havilland Aeroplane Hire Service in 1922 and the other A T & T examples were broken up after the airline ceased operation.

Handley Page 0/400, 0/7and 0/10

In addition to the enclosed D.H.4As, two or three of the larger Handley Page 0/400 twin-engined bombers were converted for the carriage of VIPs by the Royal Air Force. These were painted with aluminium dope in place of the standard dark green, had windows in their fuselage sides and bore on their noses the letters HMAL (H M Air Liner) above the aeroplane's individual name. D8326 was named Silver Star, one was called Great Britain and there may have been Silver Queen but there is no photographic evidence. Silver Star had a cabin with six seats arranged round a circular table. The seats had chintz-covered cushions and the large window had curtains. Great Britain had eight seats which were leather covered and there were sideways-facing seats along the port side of the fuselage. There were four small square windows in each side.

When Handley Page Transport began London-Paris services on 2 September 1919, the airline used 0/400s and 0/7s. The 0/400s retained their internal fuel tanks and seats were fitted behind and in front of the tanks. Seven passengers could be carried in the cabin and a further two in the nose-gunner's open

View forward in the cabin of the prototype Vimy Commercial.

Looking aft in another Vimy Commercial, with improved seating.

F-ADDT Languedoc and F-AECU Normandie (beyond) were two of Air Union's Salmson-powered Farman Goliaths. They are seen at Le Bourget with an Air Union Blériot Spad 33 and a KLM Fokker F.III. (Air France)

A very early example of a Farman Goliath equipped to carry passengers. The open cockpit, between main and front cabins, can be seen at top centre.

cockpit. The O/7 (also known as the O/700) was a much improved passenger aircraft. The fuel tanks were transferred to the nacelles aft of the engines and the fuselage cross-bracing tie-rods were replaced by diagonal tubular struts from the upper longerons to the centre of the floor and eight windows were installed in each side. One report stated that there were five rows of facing wicker seats but a Handley Page drawing shows the main cabin with three seats on the port side and four on the starboard side with the two front seats aft-facing. There was a baggage hold in front of the main cabin and forward of that was a four-seat cabin. Two passengers could still be carried in the open nose-cockpit. There was a lavatory right aft. A 4ft 6in by 2ft 2in door on the port side gave access but boarding still required a ladder. The main cabin was 15ft 6in long, 6ft 2in high and had a maximum width of 4ft 8in. The front cabin was 7ft 3in long, 5ft 5in high at the rear and 3ft 7in at the front with a maximum width of 4ft 6in. The nose-cockpit, with two side-by-side seats, had a maximum width of 2ft 9in. The O/7's windows measured 1ft 8in by 1ft 5in.

A much-improved Farman Goliath interior. (Air France)

The front cabin of the Farman Goliath provided passengers with a fine view but not protection in this kind of incident.

Some O/400s were converted to freighters as O/11s, these were later modified to O/7 standard and designated O/10s. A Number of these large Handley Page aircraft were exported, mainly to China.

Vickers Vimy Commercial

In 1917 the Air Board asked Vickers to produce a bomber to the same specification as the Handley Page O/400. It was designed in four months but was too late to see active service. Named Vimy, this large twin-engined biplane gained a place in history when one made the first direct air crossing of the North Atlantic and another made the first flight from England to Australia. Two others were engaged in the first flight to South Africa.

Following the Armistice, Vickers looked to the civil market and decided to build a civil Vimy with new fuselage for passenger carriage. This was the Vimy Commercial which first flew on 13 April 1919. This aircraft retained the wings, tail unit, engines and undercarriage of the Vimy but had a deep forward fuselage of upright oval cross-section completely clear of any internal structural obstructions. It was a plywood monocoque structure of generous proportions. There were circular windows in each side and on the port side at the rear of the cabin was a narrow opening closed by a roller blind.

The cabin windows were soon changed to rectangular shape and a new entry door was fitted. This was hinged at its base and incorporated steps — the first example of airstairs.

The Goliaths of the Farman Line on Paris — Amsterdam services had much superior cabins including curtains, painted panels of Dutch landscapes and a route map.

A Farman Goliath with sleeping berths. This may have been the aircraft used in June 1922 by Grands Express Aériens on the first civil airline night flights between Paris and London.

There was seating for ten passengers and various layouts are known. One Vimy Commercial had individual wicker seats each side of a narrow aisle and a single upholstered seat against the rear bulkhead and another had leather upholstered seats with arm rests and a full-width seat against the rear bulkhead. The cabin had electric light but there does not appear to have been any soundproofing.

Forty Vimy Commercials were exported to China but it has been reported that many of these were never uncrated. The best known example was G-EASI City of London delivered to S Instone & Co which became The Instone Air Line and in 1924 a constituent of Imperial Airways to which the Vimy Commercial passed. It became a freighter and was finally withdrawn in 1926 but its cabin was used as a summer house until 1935.

The Vimy Commercial had a maximum speed of 98mph and a range of 450 miles.

France

Like the United Kingdom, France had to rely on converted military aircraft or adaptations of military designs when beginning its airline operations. One of these types was used in considerable numbers and another was destined to remain in service for a decade.

Breguet 14

The Bregeut 14 was a single-engined two-seat biplane which first flew in November 1916 and there were two main versions, the A.2 reconnaissance type and the B.2 day-bomber. More than 5,000 were built.

Unlike the British aircraft described which were of wooden construction, the Breguets had duralumin and steel fabric-covered structures.

Passengers dressed for the joys of flight in one of Rumpler Luftverkehr's Rumpler C Is.

Immediately after the war, Lignes Aériennes Latécoère began development of an air route linking France and South America, and the airline employed more than a hundred Breguet 14s but these still had open cockpits.

CMA — Cie des Messageries Aériennes began a Paris-London service on 16 September 1919 and on this and other services employed at least 24 Breguet 14s, several of which had small passenger cabins ahead of the pilot's open cockpit.

In the first version the decking above the upper

longerons was deepened, a two-seat cabin installed and windows added. This was the 14T2 Salon and it had an entrance door in the starboard side between the upper and lower longerons. Other versions had three and four passenger seats. Ladders were still used for access.

Passengers in the open cockpit of one of Deutsche Luft-Reederei's L.V.G. C VIs. (Lufthansa)

Farman Goliath

The important French transport aeroplane was designed as a bomber but it was too late to take part in the war and was converted to a passenger type with two cabins - the rear cabin having seats for eight passengers and the front cabin four seats. On the port side between the cabins was the open cockpit.

About 60 Goliaths were used on scheduled air services. There were several versions with a range of different engines and the standard of the interiors was constantly improved. The first Goliath had structural members running diagonally through the cabin, very rudimentary wicker seats and the cockpit was not blanked off from the passenger cabins. Windows ran the full length of the cabins and round the nose.

Production aircraft had bulkheads fore and aft of the cockpit, much improved seats with tables and there were light-luggage racks. Seats were fixed to the floor.

Goliaths operated by the Farman Line were even further improved although they had wicker seats. They were used on Paris-Amsterdam services and the front bulkhead to the main cabin bore a route map, tulips were painted above the curtained windows and there was a painting of a Dutch landscape with windmills.

At least one Goliath, possibly of Air Union, had four sleeping berths in two tiers but they may never have been used. Goliaths were operated by several French airlines, by SNETA and Sabena in Belgium, CSA in Czechoslovakia and LARES in Romania. A few were used in South America and at least six were built in Czechoslovakia. Some Goliaths remained in service until 1933. A Goliath of Grands Express Aériens was involved in the first collision of transport aircraft when, on 7 April 1922, it collided with a Daimler Airway de Havilland D.H.18 near Poix.

Germany

Deutsche Luft-Reederei (DLR) began operating German domestic air services in February 1919 with a fleet mainly consisting of wartime two-seat open-cockpit biplanes. A small number were later equipped with small and rather primitive cabins.

This Breguet 14T of CMA (Cie des Messageries Aériennes) was a two-seat-cabin version converted from a Breguet 14A.2 reconnaissance aircraft.

An early example of German air transport. Passengers in a Deutsche Luft-Reederei A.E.G. GV, appear to be taking courage before departure or celebrating their safe arrival. (Lufthansa)

This D.F.W. P1 was probably built in 1919 and said to have four seats. At least it had curtains.

Passengers board the inaugural DDL — Danish Air Lines' flight on 7 August 1920. The route was Copenhagen — Warnemünde and the aeroplane a Friedrichshafen FF49 seaplane. The ladder was an essential piece of airport equipment in the early days of air transport. (DDL)

L.V.G. C VI

L.V.G. C VIs formed a large part of the DLR fleet, and were used by a number of smaller German airlines and also by SNETA in Belgium.

The C VI was a wood and fabric single-engined reconnaissance biplane of which approximately 1,000 had been built by the end of the war. In airline service, the rear cockpit accommodated two passengers and the only concession to their comfort was the loan of flying clothing.

Rumpler Biplanes

Rumpler Luftverkehr began operating Augsburg-Munich-Nuremberg-Leipzig services in 1919 and employed a number of C I and C IV single-engined two-seat biplanes.

In an effort to provide some passenger comfort a number of C Is were given two-seat cabins. These were raised bent-ply structures with Triplex windows. Seats were side-by-side and there was a speaking tube for communication with the pilot. In this form the C I was designated Rumpler 5A2.

A.E.G. G V

Although DLR's fleet were mostly open-cockpit single-engined biplanes the airline did have a number of twin-engined aircraft. Among these were some A.E.G. G V bombers which had been built too late to go into military service. The large 27.3m (89ft 6in) span biplanes had an open cockpit for two crew and an aft cockpit which could accommodate six passengers. A few A.E.G. G Vs were produced with six-seat cabins on two levels and with a lavatory aft of the cabins.

Friedrichshafen FF 49

DLR and some other German airlines operated a number of Friedrichshafen FF 49 single-engined two-seat twin-float reconnaissance seaplanes on coastal services and one of this type operated DDL — Danish Air Lines' first service on 7 August 1920. A few cabin versions of the FF 49 were produced. The two passengers travelled in the large rear cockpit.

CHAPTER FOUR
FIRST POST-WAR AIRLINERS

The Handley Page W.8 appeared at the end of 1919 and led to a family of twin and three-engined biplanes.

United Kingdom

Handley Page began producing a series of successors to its post-war conversions and de Havilland, as successor to Airco, designed and built new transport aeroplanes. The Bristol Aeroplane Company also began airliner design. The most important new airliners of the period were produced by Fokker in The Netherlands and Junkers in Germany and are the subject of Chapter Five.

Handley Page W.8

Handley Page realised that internal structure impeding the cabin had to be overcome and

The cabin of the Handley Page W.8. Curtains and floor carpet added to comfort but the candle-holders raise questions.

The cabin of a W.8b with fourteen seats, electric light and a BP route map on the ceiling.

Three Handley Page W.8bs were operated by Handley Page Transport from 1922 and they passed to Imperial Airways in 1924. One W.8b was built for Sabena and another three were built in Belgium.

designed the W.8 with completely unobstructed cabin. The W.8 was a wood and fabric twin-engined biplane with 22ft long cabin which had a width of 4ft 6in and a height of 7ft. There were forward-facing seats on each side and sixteen could be provided with a narrow aisle although the aircraft was only certificated for twelve passengers. There were eight circular curtained windows on each side. Brackets looking like candle holders presumably held cabin lights and there was a clock on the front bulkhead. The W.8 began operating Handley Page Transport London-Paris services in October 1921 but only remained in service for a short time.

Handley Page W.8b

The W.8b was an improved version of the W.8. The type first flew on 21 April 1922, three were built and operated by Handley Page Transport before passing to Imperial Airways as Princess Mary, Prince George and Prince Henry. The W.8b's interior was much better than the W.8's. Windows ran the full length on each side, there were fourteen cushioned wicker seats and above the windows were light-luggage racks. The cabin had electric ceiling lights. The cabin length including the lavatory was 19ft, and it was 4ft 8in wide and 6ft high. A fourth W.8b was built for Sabena and a further three were built in Belgium.

Handley Page W.10

Handley Page built a three-engined version of the W.8 and also a three-engined W.9 but the last of the series were four examples of the W.10 which entered service with Imperial Airways in 1926. W.10s had sixteen seats in a cabin 22ft 10in long, 4ft 5in wide and 5ft 9in high. Two of the four were lost in the English Channel.

The Handley Page W.10 G-EBMR City of Pretoria at Croydon. Imperial Airways had four W.10s and two of them were lost in the English Channel.

The D.H.18 was the first airliner produced by de Havilland Aircraft.

de Havilland D.H.18

Although generally known as the de Havilland D.H.18, it was in fact the last Airco type and some of them were built by the new de Havilland Aircraft Co. Before designing the D.H.18 Geoffrey de Havilland had produced drawings for the D.H.17 which was to have been a sixteen-passenger twin-engined biplane with semi-retractable undercarriage but it was not built.

The D.H.18 was a single-engined biplane with seats for eight passengers. There were two aft-facing seats against the front bulkhead and two forward-facing seats against the rear bulkhead. Between the bulkhead seats were two forward-facing seats on the starboard side and two rearward-facing on the port side. Access must have been difficult because there was no aisle. There were forward and aft escape hatches and the ply-covered fuselage was watertight in case of a forced alighting at sea. Only six D.H.18s were built and they served with Aircraft Transport & Travel, The Instone Air Line and the Daimler Airway.

The cabin area of the D.H.18.

de Havilland D.H.34

The D.H.34 was a wooden single-engined biplane with seats for eight or nine passengers. The

ABOVE: The interior of an Imperial Airways de Havilland D.H.34.
LEFT: The mixture of forward-and rearward-facing seats in the D.H.18.

Instone Air Line's de Havilland D.H.34 City of New York, at Brussels.

The view forward from the Bristol Pullman's rear cabin. (Bristol Aeroplane Company)

Looking aft from the Bristol Pullman's front cabin.

first example flew for the first time on 26 March 1922 and fully-certificated it was delivered to the Daimler Airway five days later. Two entered service on 2 April with Daimler and The Instone Air Line on the London-Paris route.

The cabin had a length of 15ft 5in and was about 4ft wide. There were four seats each side and a single seat aft of the lavatory door. The windows had curtains, there were light-luggage racks and heating and ventilation could be controlled from within the cabin. On Daimler aircraft a cabin boy was carried.

Eleven or twelve D.H.34s were built, seven Instone and Daimler aircraft passed to Imperial Airways and one example was exported to the Soviet Union.

For some long forgotten reason, the passenger door was in the starboard side which was very unusual for a British aeroplane.

Bristol Pullman

The large Bristol Pullman passenger version of the Braemar bomber flew in May 1920 and was equipped to carry fourteen passengers. It was an unusual design being a triplane with two tractor and two pusher engines mounted above the centre wing.

The Pullman never went into service and is only included in this book because of its high standard of passenger accommodation. As can be seen in the illustration the Pullman had generous ceiling height, curtained windows and electric light. There were ventilators in the ceiling.

ABOVE: The only triplane in the book. The Bristol Pullman of 1920 was developed from the Braemar bomber and equipped to carry fourteen passengers but never went into service. (Bristol Aeroplane Company)

Bristol Ten-seater

The Bristol Ten-seater first flew in June 1921, but only three were built and they saw limited service, partly as freighters. As with the Pullman, the Ten-seater is included in this book because of its attractive interior. It had nine passenger seats, the tenth person of its designation was the pilot. The purpose of the circular objects on the walls is not known.

Germany
Dornier Komet

This very small single-engined type first produced in 1921 can hardly be called an airliner but it was to lead to a family of aircraft.

The Dornier Do C III Komet, later known as Komet I, unlike the British aircraft described was a monoplane and of all-metal construction. It only had a length of 9.5m (31ft 2in) but contained a four-seat cabin with two forward-facing and two backward-facing seats. There were three large windows in each side.

Deutsche Luft-Reederei used Komet Is and one of them was the first German aeroplane to land in England after the 1914-18 war.

A more powerful Komet II was produced in 1922 and a much changed six-passenger Komet III in 1924, with some built in Japan.

ABOVE: An early German transport aeroplane, the Dornier Do C III Komet of 1921. There were two forward-and two rearward-facing seats. The small size of the Komet can be judged from the passenger seen through the cabin window. (Swissair)

BELOW: Much larger but still a Komet was the six-pasenger Komet III of 1924.

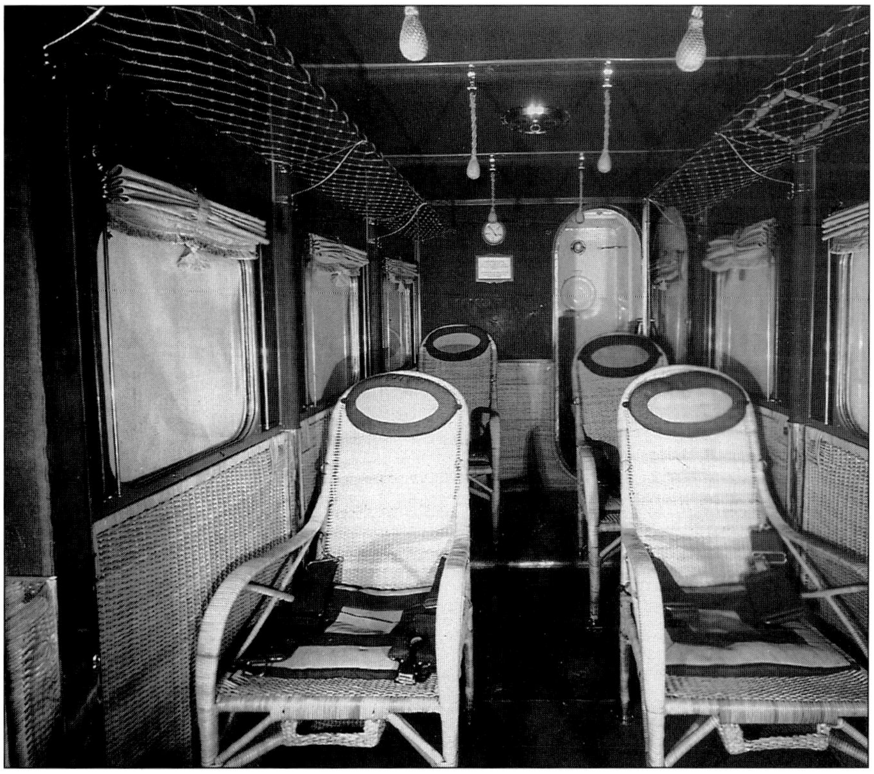

The Bristol Type 75A Ten-seater after conversion to freighter. (Bristol Aeroplane Company)

The well-appointed cabin of a Bristol Ten-seater of 1921. (Bristol Aeroplane Company)

Back to the ladder. Entry to the Fokker F.II required a helping hand. (Lufthansa)

The German-built Fokker-Grulich F.II was used by several German airlines and D-756 Dievenov is seen in service with Luft Hansa. (Lufthansa)

CHAPTER FIVE
FOKKER AND JUNKERS

A Fokker F.III of the German-Russian airline Deruluft. (Lufthansa)

With the exception of the Bristol Pullman triplane and Dornier Komet, all aircraft so far described were biplanes of mainly wooden construction with fabric or plywood and fabric covering.

Fokker in Germany and The Netherlands and Junkers in Germany both produced monoplane airliners. Junkers pioneered all-metal construction, Fokker employed metal structured fuselages combined with wooden wings. Both companies produced families of aircraft with such structures and they played a major role in the development of air transport in Europe and a number of other areas.

Fokker F.II and F.III

During the 1914-18 war Fokker had a factory in Germany and produced some of the best fighter aircraft for the German military air services, the last type being the D VIII parasol-wing monoplanes. When the war ended, Fokker turned to commercial aircraft and these were influenced by the D VIII design. Work began on the V.44 (F.I) which was to carry six passengers in open cockpits. Fortunately this aeroplane was abandoned and the cabin V.45 (F.II) was built instead.

Like all Fokker transports between the wars, the F.II had a wire-braced welded steel-tube fuselage which was fabric-covered and a ply-covered all-wood one-piece wing mounted above the cabin. The original engine was a 185hp BMW. The enclosed cabin had three windows in each side, a door in the port side and seats for four passengers. A fifth passenger could be carried beside the pilot in an open cockpit between the cabin and the engine.

The F.II first flew in October 1919 and was acquired by KLM. A few F.IIs were built by Fokker in The Netherlands and some were built in Germany as Fokker-Grulich F.IIs. As far as can be discovered there were 24 F.IIs, they served a number of airlines and some remained in service until the mid-1930s.

The F.II was followed in April 1921 by the very similar but more powerful and slightly larger F.III which had a wider fuselage and a five-seat cabin. The pilot still sat in an open cockpit but beside the engine. The cabin had three windows each side and the seating consisted of a full-width seat against the rear bulkhead, a forward-facing armchair on the port side and a similar inward-facing seat on the starboard side. All seats were upholstered and the single seats could be anchored to the floor.

The cabin of a Fokker F.III. The keys suggest the seats could be anchored to the floor. (KLM)

ABOVE: Ski-equipped Junkers-F 13s of Aero O/Y (Now Finnair) at Helsinki.
BELOW: The four-seat cabin of an Ad Astra Aero Junkers-F 13. It has curtains, electric light and ventilation.

BELOW: The Junkers-F 13 has always been regarded as a four-passenger aircraft but this view shows that a fifth passengr could be carried. Of great interest are the seat-belts. These were not required in British aircraft until October 1937. (Lufthansa)

Twelve F.IIIs were built for KLM and as with the F.IIs, F.IIIs were built in Germany as Fokker-Grulich F.IIIs. There seems to have been a total of 76.

The F.II and F.III led to a whole family of Fokker airliners all with fabric-covered welded steel-tube fuselage and one-piece high-mounted wooden wings. The family included the several versions of the outstanding F.VII series, one of which made the first trans-Pacific flight from the USA to Australia.

Junkers-F 13

The Junkers approach was different to that of Fokker. Its transport aeroplanes were low-wing cantilever monoplanes of all-metal construction with corrugated metal skins to all surfaces.

Like Fokker, Junkers' first transport aeroplane design was for an aircraft with only semi-enclosed passenger accommodation. But this J 12 was abandoned and instead the J 13 was built with a completely enclosed cabin for four passengers. The semi-enclosed cockpit had two seats.

The J 13, later designated F 13, was a very important aeroplane. It had a multi-spar wing with duralumin tubular spars braced to form an extremely strong girder. The entire structure was very strong, had a long life and was virtually impervious to weather. So successful was this form of construction that it was continued up to the Junkers-Ju 52s of the 1930s.

More than 320 F 13s of some 60 or 70 variants were built, they were widely used in Europe and saw service in many parts of the world. The F 13 first flew on 25 June 1919 and was an outstandingly advanced aeroplane for its time.

The cabin had two full-width seats in front of the rear bulkhead and two individual seats in the front of the cabin. The F 13 has always been regarded as a four-passenger aircraft but there is photographic evidence that the rear seat could accommodate three passengers.

There were three windows in each side, a door in the port side and electric light in the roof. The F 13 may have been the first passenger aeroplane to have seat-belts.

Although most F 13s were orthodox landplanes, the type was also fitted with skis for winter and several were twin-float seaplanes. Entry was via steps in the wing root.

CHAPTER SIX
1920s - EUROPE

Imperial Airways' Armstrong Whitworth Argosy City of Glasgow. (Charles E Brown)

In spite of the adoption of the monoplanes and metal construction by Fokker and Junkers many of the European airliners of the 1920s were biplanes made of wood or at best of wood with some use of metal, and average passenger capacity seldom exceeded twenty. Two British types must be mentioned, both being three-engined biplanes, one of wood and the other of mixed construction.

United Kingdom
Armstrong Whitworth Argosy

The Argosy was an important aeroplane although only seven were built. It was a large three-engined biplane originally designed as a Middle East aircraft but it was considered suitable for European routes and three were ordered in 1925. The type first flew in March 1926 and went into service with Imperial Airways between London and Paris that July.

ABOVE: The cabin of a twenty-seat Argosy. There is an emergency exit in the roof and a sick bowl beneath the seats.
BELOW: An eighteen-seat Argosy, with Steward and buffet, as used on the Silver Wing London — Paris service. This one has improved seats and life-jackets above the windows.

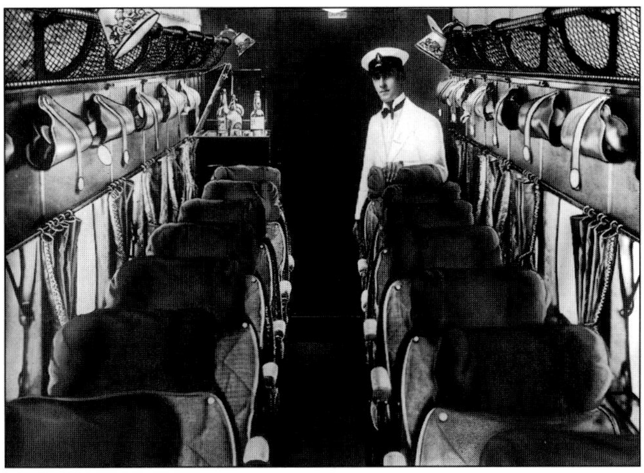

The fuselage was a fabric-covered steel-tube structure but the wings were of wood with fabric covering. The cabin was 29ft long, 4ft 6in wide and 6ft 2in high. There was a wooden floor and the walls were lined with fabric. There were twenty forward-facing wicker seats, opening windows stretched the length of the cabin and there were light-luggage racks above the windows. There were containers for sick bowls beneath the seats and also electric lighting.

In May 1927 Imperial Airways introduced the Silver Wing London-Paris service. For this service, the

A de Havilland D.H.66 Hercules with the port engine shut-down. The first examples had open cockpits. (Flight)

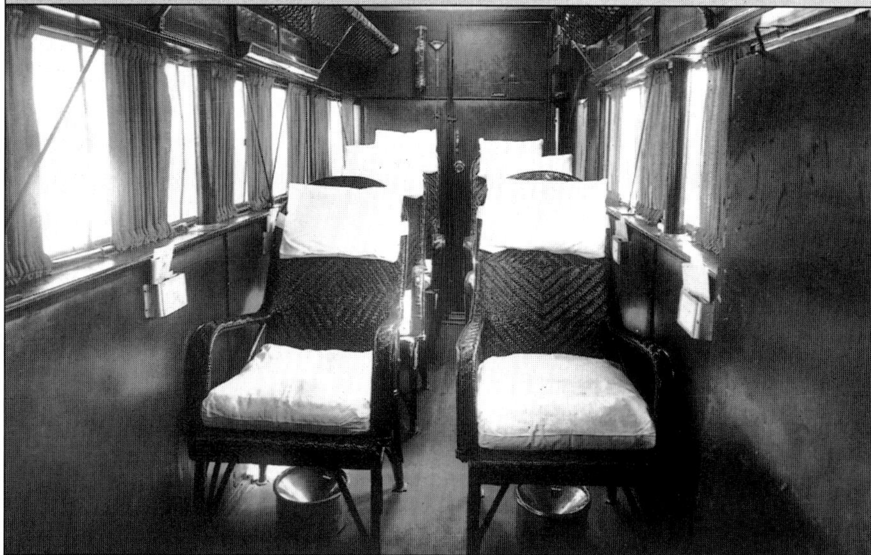

LEFT: The cabin of an Imperial Airways' D.H.66 Hercules. It has a fire-extinguisher, not then a requirement and there are sick bowls beneath the seats.

BELOW LEFT: The six-passenger cabin of a Blériot Spad 56/4. The windows could be opened and there was an emergency exit in the ceiling.

BOTTOM: A CIDNA Blériot Spad 56/4. There were more than 100 similar aircraft built.

Argosies were given improved interiors with eighteen cushioned seats with armrests, curtained windows and at the rear on the starboard side a buffet. A steward was carried for the first time on British aircraft. The Argosy had a lavatory behind the passenger cabin.

From October 1927 Imperial Airways used the Argosy on the London-Paris route with a 2hr 30min schedule and £9 return fare. The services had a steward and a bar, and operated second-class services with the surviving Handley Pages, without steward, to a 2hr 50min schedule and a £7.10 (£7.50p) return fare.

A further three, improved aircraft, were ordered in 1928 and in early 1931 two were transferred to Cairo to operate the Cairo-Khartoum sector when the first stages of the service to Cape Town began. The last Argosy was used for joy-riding at Blackpool by United Airways in the summer of 1936.

Among the ugliest of early airliners was the Farman Jabiru (Stork), which was built in a variety of forms with one, two, three and four engines. This is the Farman Line's F.3X Jabiru with four Hispano-Suiza engines.

de Havilland D.H.66 Hercules

From June 1921, the Royal Air Force operated the Desert Air Mail route between Cairo and Baghdad but in 1925 the Air Ministry made an agreement for Imperial Airways to take over this service and it became the first stage of the eventual United Kingdom-India route.

In order to take up the operation Imperial Airways issued a specification for a new aircraft. This was produced as the de Havilland D.H.66 Hercules three-engined biplane which began operation between Basra, Baghdad and Cairo on 7 January 1927. The service was timed to connect with P & O mail ships linking Port Said and Marseilles.

The Hercules had a tubular steel fuselage frame and the passenger accommodation was a large plywood box inserted into the forward fuselage. It had windows along each side with light-luggage racks above the windows. There were electric lights in the ceiling and a fire extinguisher on the rear bulkhead. It is difficult to determine how many passengers could be carried. The accompanying photograph shows six armchairs but Imperial Airways' documents state that the Hercules had accommodation for eight passengers and a de Havilland work's drawing dated December 1927 shows ten seats, staggered so that the starboard seats were aligned between the port side seats and slightly further aft.

These first Empire routes aircraft worked through to Karachi when the England-India service was inaugurated in March 1929 and operated the Kisumu-Cape Town section of the England-South Africa route when that was opened in January 1932.

In 1929, West Australian Airways acquired four Hercules to operate Perth-Adelaide services. These aircraft had a number of modifications including closed cockpits and had seating for fourteen passengers. Two of the Australian aircraft passed to Imperial Airways and the earlier aircraft were modified to the later standard.

The interior of a Farman Jabiru showing main and nose cabins.

The first example of the 24-passenger Farman F. 180 Oiseau Bleu (Bluebird) was delivered to the Farman Line in 1928.

France
Blériot Spad 33 Series

In 1920 Blériot-Aéronautique built the first of a family of very successful small single-engined passenger biplanes which remained in production for a decade. They had wooden monocoque fuselages.

The first of the series was the Blériot Spad 33 which cruised at 160km/h (99.4mph) but some of the models had a maximum speed of 200km/h (124.2mph).

In the Spad 33, 46 and 50 the passenger cabin was between the wings ahead of the pilot's open cockpit but in the 56 (illustrated), the passenger cabin was lengthened and placed further aft with the cockpit ahead of the cabin.

There were more than a hundred Blériot Spads and they were operated by a number of French airlines including CIDNA (Compagnie Internationale de Navigation Aérienne) which used them on its trans-Europe services which stretched as far as Constantinople, now Istanbul.

Most of the series had circular cabin windows which could be opened and an additional roof window which served as an emergency exit. The cabin had four individual leather upholstered seats, with seat-belts, and there was a full-width bench seat against the rear bulkhead.

The Farman F.180 with some sleeping accommodation but it is unlikely this was ever used. (Air France)

Farman F.121 Jabiru (Stork)

The very ugly Farman F.3X Jabiru was first built in 1923. It had a wooden wing and mixed construction fuselage which at the nose was almost 3m (10ft) deep and contained a two-seat cabin. Aft of this cabin there was a section with semi-bulkheads which had a single seat on the starboard side, a fuel tank on the port side and in the top the open cockpit. The main rear cabin had six seats. Windows ran the length of the cabin and round the nose. There were electric roof lights and in the main cabin overhead light-luggage racks. All the wicker seats except the midship single seats were canted slightly inward.

Luft Hansa's Junkers-G 24 Hestia at Tempelhof Airport, Berlin. The G 24 first flew in mid-1925 and was a more powerful version of the G 23 of 1924. The G 23 and G 24 were the first European three-engined all-metal airliners.

The four 180hp Hispano Suiza engines were mounted in tandem pairs at the end of stub wings. Cruising speed was 175km/h (108.7mph).

The Farman Line had four F.3X Jabirus on its Paris-Brussels-Amsterdam route and DDL — Danish Air Lines used the type on Copenhagen-Hamburg-Cologne services. Two of DDL's Jabirus were French-built and two were built in Copenhagen.

CIDNA had four F.4X Jabirus of hideous appearance. These had no forward cabin but the nose held three water-cooled radial engines. There must have been high noise and vibration levels. They were all withdrawn from service after a short time. There was also a single-engined version for eight passengers. This was the F.170 Jabiru operated by the Farman Line on Paris-Berlin services.

ABOVE: Passengers boarding a Junkers-G 24 at Tempelhof.
BELOW: View forward in the cabin of a Junkers-G 23. (Junkers)

Farman F.180 Oiseau Bleu (Bluebird)

This was a large twin-engined biplane built for an attempted North Atlantic flight which was instead delivered to the Farman Line in 1928. It was of wooden construction and the upright oval-

Luft Hansa's Focke-Wulf A 17 Möwe Leer. Luft Hansa used eleven of these eight-passenger aircraft on domestic routes. (Swissair)

The very clean interior of the Junkers-G 23. View looking aft.

section fuselage was ply-covered inside and out. The cabin had 24 seats, three across, and seven large rectangular windows each side which could be opened. There were at least three F.180s and at least one was fitted with sleeping accommodation for twelve passengers. The illustration shows the sleeper version with four berths in two tiers on the port side at the front of the cabin with long inward-facing seats on the starboard side. Aft there were forward-facing seats with single seats to starboard and double seats on the port side. There were overhead net luggage racks and a patterned floor carpet.

A Deruluft Dornier Merkur. Merkurs began to appear in 1925,some were newly-built and others converted from Komet IIIs. (Lufthansa)

One of the most important airliners of the 1920s and 1930s was the Fokker F.VIIb/3m. This one was operated by Japan Air Transport.

Farman F.300

In 1929 Farman began production of single- and three-engined monoplanes under the designation F.300. The cabins had eight upholstered seats, four on each side and there were full-length luggage racks. Of interest were the notices above the windows in three languages saying 'No Smoking' and others stating that it was forbidden to throw anything out of the windows.

Farman F.300s were operated by the Farman Line under the class name l'Etoile d'Argent (The Silver Star) and when Air France was founded, the new airline took over fifteen of the series. Six were used by Air Orient with four based at Damascus.

Germany
Junkers-G 23 and G 24

The Junkers-G 23 which first flew in 1924 and the G 24 of mid-1925

A KLM Fokker F.VIIb/3m as furnished for the Far East route. (KLM)

The forward part of the cabin of a Sabena Fokker F. VIIb/3m.

A Fokker F.XII of Crilly Airways. This one was originally delivered to KLM in 1931 as PH-AIE. (KLM

The Fokker F.XII as laid out for KLM's Amsterdam — Batavia (now Jakarta) route. (KLM)

A KLM Fokker F.XII configured for European services.

The rear part of the cabin of ABA — Swedish Air Lines' Fokker F.XII. Värmland.

both deserve a place in history. They were the first European multi-engined all-metal monoplane airliners and they set new standards in passenger comfort, reliability and safety. Their structures continued the pattern established with the single-engined F 13 of 1919 and like the F 13 had corrugated metal skins.

G 23s were put into production in Sweden and ABA - Swedish Air Lines had four. Ad Astra Aero in Switzerland had four and one, as a seaplane, was used by Aero Lloyd in Poland before being converted to a G 24.

The G 24 was heavier and more powerful and was built in Germany and Sweden. Total production of the two types was about 60.

During 1926 Luft Hansa used two G 24s for a 20,000km (12,427 miles) flight from Berlin to Peking (now Beijing) and back to assess the possibility of an air route to the Far East via Siberia.

After passenger service, several G 24s were converted to F 24s as single-engined aircraft. Normal accommodation was for nine passengers in leather upholstered seats. There was a large openable window beside each seat. There were net luggage racks, electric roof lights and aft of the cabin a wash room and lavatory. A door in the front bulkhead gave access to the cockpit and it contained a window which gave passengers a view of the controls. According to Junkers, a page-boy was carried to give any required assistance to passengers.

Focke-Wulf A 17 Möwe (Gull)

During the 1920s several German aircraft manufacturers produced single-engined monoplanes seating eight-to-ten

Air Union's red and gold Lioré et Olivier 213 F-AIZN at Le Bourget. The type operated under the fleet name le Rayon d'Or. (The Golden Ray)

passengers. Typical of these aircraft was the eight-passenger Focke-Wulf A 17 Möwe (Gull).

Dornier Komet III and Merkur (Mercury)

The closely-related Dornier Komet III and Merkur were also single-engined high-wing monoplanes. The Komet III was a completely new aeroplane and bore virtually no resemblance to the Komet I and II described in Chapter Four.

The Komet III first flew at the end of 1924 and had accommodation for six passengers. They were used in Germany by Deutsche Aero Lloyd, DDL in Denmark had one and the type was put into production by Kawasaki in Japan. Six German-built aircraft served in the Ukraine.

In 1925 Dornier produced the very similar but more powerful Merkur and some Komet IIIs were brought to Merkur standard.

Luft Hansa had twenty-two Merkurs on Berlin-Königsberg-Moscow night services and Deruluft had at least nine on Königsberg — Moscow services. Some Merkur seaplanes operated in Brazil and in December 1926 -

Looking aft in Air Union's Lioré et Olivier 212 restaurant aircraft. (Air France)

Standard layout of a Lioré et Olivier 213. (Air France)

February 1927, the well known Swiss pilot Walter Mittelholzer flew a Merkur seaplane from Zürich to Cape Town in 100 flying hours. A photograph of a Komet III cabin appears in Chapter Four.

The Netherlands
Fokker F.VII series

Truly outstanding transport aeroplanes of the 1920s were the Fokker F.VII series which were produced in large numbers and numerous versions.

The cabin of one of CIDNA's eight five/eight-seat Bernard 190Ts which was used on the trans-Europe route from 1929. (Air France)

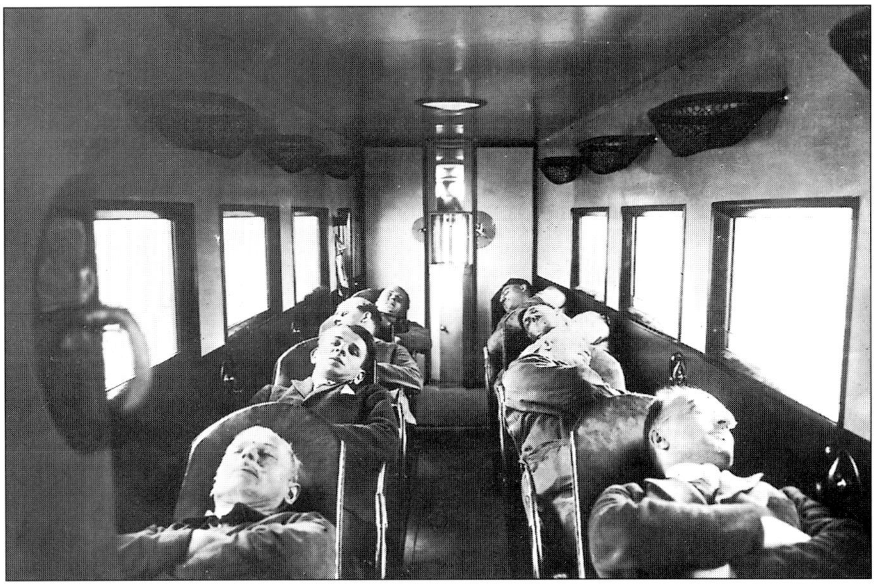

The cabin of one of the Albatros L 73s built for night operations by Luft Hansa. They had eight reclining seats and it was reported that four of the seats could be converted to two sleeping berths.

The F.VII was designed to operate KLM's forthcoming services over the very long route from Amsterdam to Batavia in the Netherlands East Indies (now Indonesia), and one of the five examples made a flight over the route in 1924.

Like the F.II and F.III, the F.VII was a single-engined high-wing monoplane with fabric-covered welded-steel-tube fuselage and ply-covered one-piece all-wood wing. The engines were water-cooled.

A modified version, the F.VIIa, mostly with air-cooled engines, was put into production, 40 were built and in 1927 one made the first passenger-carrying flight to Batavia. It had eight passenger seats. F.VIIas were built with one and three engines.

The most successful of the series was the F.VIIb-3m produced in 1928. Of similar layout and construction and still with eight seats the type remained in production until 1932 with a total of 145. Some were assembled in the United States, 28 were built in Belgium by SABCA for Sabena, others were built in Czechoslovakia, Poland and Italy and fourteen were built in the United Kingdom as Avro Tens with five going to Australian National Airways. The Sabena aircraft had nine seats and a few F.VIIb-3ms had accommodation for ten passengers.

Apart from their airline work these aircraft made some outstanding long-distance flights. One made the first flight from the United States to Hawaii, Admiral Byrd made the first aeroplane flight over the North Pole in one and Charles Kingsford Smith and crew in the Southern Cross made the first air crossing of the Pacific, from California to Brisbane.

Some single-engined F.VIIs were converted to have three engines.

Fokker F.XII

KLM required a larger aeroplane to replace the eight F.VIIb-3ms used on the Batavia route and to meet this requirement Fokker produced the larger F.XII of the same configuration. Although it did not actually fly until 1931 it is convenient to mention it here. The F.XII had sixteen seats when operating in Europe but for the Far East route it was equipped with four sleeper seats.

Luft Hansa's Junkers-G 31 Westmark at Schiphol Airport, Amsterdam. In the background are a Swedish Air Lines' Junkers-G 23 and Luft Hansa Fokker-Grulich F.III. (KLM)

Eleven F.XIIs were built in The Netherlands and another two in Denmark. AB Aerotransport (Swedish Air Lines) had one of the Dutch-built aircraft and it flew until 1946 when it was destroyed in a hangar fire.

For comparison with today's airliners, the maximum weight of the F.XII was 7,750kg (17,085lb) and it cruised at 127.3mph.

France
Bernard 190T

The eight Bernard 190T single-engined five-to-seven seat monoplanes used by CIDNA on its trans-Europe route from 1929 were of no special significance but the type is included here for its jazzy interior. Unfortunately the colours are not known.

Lioré et Olivier 21

After operating a few four-engined Blériot biplanes Air Union acquired, from 1926, a fleet of Lioré et Olivier 21 twin-engined biplanes. These were painted red and gold and operated under the fleet name le'Rayon d'Or (The Golden Ray).

The wings were of metal construction with fabric covering and the fuselages of mixed construction.

The main cabin had twelve seats and there was a six-passenger cabin in the nose. The two cabins were connected by a passageway which also gave access to the cockpit.

The later LeO 213 had the forward cabin replaced by a baggage hold and two LeO 212s were equipped as restaurant aircraft with seats for twelve and a bar. Seats were on the port side arranged in facing pairs across tables. These restaurant aircraft, introduced in July 1927, had along the fuselage the text 'Avion Restaurant Londres-Paris-Marseille-Tunis' but they probably only worked between London and Paris in competition with Imperial Airways' Silver Wing services.

Meal service in a Luft Hansa Junkers-G 31.

The ten-passenger Rohrbach Roland II Siegberg. The earliest Rolands had open cockpits

Germany continued
Albatros L 73

In 1926 Albatros built the L 73 for night operation by Luft Hansa. The cabin had eight seats and it was reported that four of the seats could be converted to make two sleeping berths but the photograph shows eight reclining seats. There was a lavatory and baggage hold and the large windows could be opened.

Two L 73s were used by the airline and it was stated that they operated night stages of the Berlin-Moscow route.

Junkers-G 31

In 1928 Luft Hansa began to take delivery of a fleet of Junkers-G 31s. These had the same three-engined low-wing monoplane layout and corrugated metal skins as the G 23 and G 24 but, unlike the earlier aircraft, had twin fins and rudders.

The cabins were divided into three sections by semi-bulkheads, had full-length luggage racks and the windows could be cranked open. Normal accommodation was for fifteen passengers. A steward was carried and the G 31s were probably the first Luft Hansa aircraft with meal service.

The cabin was 2.1m (6ft 10in) wide and 1.87m (6ft 2in) high and could accommodate sleeping berths or stretchers. There was a lavatory at the rear of the cabin.

Rohrbach Ro VIII Roland

From 1926 Rohrbach produced a number of Roland three-engined high-wing monoplanes of all-metal construction. The fuselage was still of rectangular box-section. The windows could be opened as seen in the photograph.

Luft Hansa employed a number of Rolands on routes which included the trans-Alpine route from Munich to Milan. Some Luft Hansa aircraft passed to 'Iberia' Cia Aérea de Transportes and at least three were used by Deruluft.

Passengers and friends shake hands through the open windows of Luft Hansa's Rohrbach Roland Watzmann. (Lufthansa)

CHAPTER SEVEN
UNITED STATES SINGLE-ENGINED TYPES

The wooden Lockheed 9 Orion was the first transport aeroplane to have a retractable undercarriage. The Orion was to have a profound influence on German aircraft design. (Swissair)

In the early years of air transport in the United States large numbers of single-engined aircraft were employed. Some of the most used are mentioned in this chapter.

Although these types played an important role in the development of air services most had to be replaced when a ban was brought in prohibiting the carriage of passengers in single-engined aeroplanes at night and over hazardous terrain.

Boeing Model 40A

In 1927 the US Post Office air mail services were turned over to private enterprise and Boeing Air Transport secured CAM 18, the route from San Francisco to Chicago which it began operating to a 20hr schedule on 1 July 1927.

To operate the service Boeing built 24 Model 40A single-engined biplanes each of which could carry two passengers in addition to mail.

In 1928 Boeing produced the modified Model 40B-4 which could carry four passengers in addition to 500lb of mail. The cabins of these aircraft, between the wings and ahead of the open cockpit, were very cramped and entry and disembarkation were not easy.

Lockheed 9 Orion

Although only 35 Orions were built the type was a most important development, being the first transport aeroplane with a retractable undercarriage and probably the first with landing flaps.

The cabin of the Lockheed Orion on display at the Verkehrshaus der Schweiz, Lucerne.

A Western Air Express Fokker F-14. It had seats for six passengers.

First flown early in 1931 the Orion was an all-wood single-engined low-wing monoplane with a very restricted cabin having five or six seats arranged up to three each side without a central aisle. Seats had to be folded to enable passengers and pilot to embark. It was the first transport aeroplane capable of 200mph. The cruising speed of one version was 205mph and the top speed 225mph.

Two Orions were bought by Swissair and to meet the competition Luft Hansa required comparable aircraft from the German aircraft industry as is related in Chapter 13 under Heinkel He 70 and He 111.

Fokker 5 Universal

Starting in 1925 Atlantic Aircraft Corporation (Fokker's United States factory) began production of the single-engined Universal designed by Robert Noorduyn who was later to design the rugged Noorduyn Norseman.

The Universal could carry four passengers and cruise at 99mph on 220hp. Forty-five were built and they were used by 31 airlines.

Fokker 8 Super Universal

The Super Universal, with six passenger seats, was a larger and more powerful development of the Universal. It first flew in 1927. With a 420hp air-cooled radial engine the Super Universal cruised at 118mph. Eighty-one Super Universals were built in the United States plus fifteen by Canadian Vickers and seventeen in Japan by Nakajima.

Fokker F-14

In 1928 production began of the Fokker F-14 which resembled other Fokker single-engined transports except that its wing was parasol mounted instead of being attached to the top of the fuselage.

The F-14 had accommodation for six

It was not easy to embark on Boeing's two-to-four-seat Model 40 series biplanes.

Fokker Model 5 four-passenger Universals were built in the USA, Canada and Japan. The aircraft seen here were operated by Nippon Koku Yuso K.K.

RIGHT: A US operated Orion with seats folded to allow embarkation of pilot and passengers. The notice on the back of the pilot's seat read 'You may enjoy a cigarette if your fellow passengers do not object'. (Courtesy R E G Davies)

passengers and cruised at 124mph. Thirty-four F-14s were built with six serving Western Canada Airways and four Western Air Express.

In August 1937 an F-14 is reported to have been the first aeroplane to have made an automatic landing.

RIGHT: The forward part of the cabin of a Fokker Super Universal.

The improved six-passenger Fokker Model 8 Super Universal. NC9724, illustrated, was one of several operated by Western Air Express. (Gordon Williams)

TOP: The eighteen-passenger Boeing Model 80A-1 with triple fins and rudders was one of ten modified Model 80As.

LEFT: The cabin of a Boeing Air Transport Boeing Model 80. Note it has three-abreast seating.

BOTTOM: A twelve-passenger Fokker F-10A (F-XA) of American Airways.

CHAPTER EIGHT
UNITED STATES DEVELOPMENT

One of the twelve-passenger Curtiss T-32 Condor II sleeper aircraft used by American Airways, later American Airlines. This one was in day configuration. (American Airlines)

In the early days of air transport Europe was ahead of the United States in passenger aircraft design, but the latter was soon to catch up and for many years took over the lead. So successful was Boeing Air Transport's San Francisco-Chicago service that there was soon a requirement for larger aeroplanes.

American Airways introduced some sleeper services and one outstanding type, the Ford Tri-Motor, played a very important part in the development of nation-wide air services.

Boeing Model 80

To provide much greater capacity and improved standards over the single-engined Boeing Model 40s Boeing designed and built the Model 80 three-engined biplane with seating for twelve passengers. Seats were in pairs on the port side with single seats to starboard. Light-luggage racks ran the length of the cabin and there were individual reading lights. The Boeing 80 first flew in August 1928, four were built, they cruised at 115mph and had a range of 545 miles.

So successful was the Boeing 80 that it was followed by ten improved eighteen-passenger Model 80As which were 10mph faster. The Boeing 80As were the first airliners to number a stewardess among the crew.

Curtiss T-32 Condor II

The Condor II first flew in January 1933 not long before the appearance of the first all-metal multi-engined monoplanes and its success as a large twin-

Part of an Eastern Air Transport Curtiss Condor cabin in day configuration. (Courtesy R E G Davies)

The Ford Tri-Motor, first built in 1926, was for many years the most important airliner in the United States. The Model 5-AT illustrated was operated by Maddux Airlines of Los Angeles from 1929. It was named The Golden Gate.

engined biplane was short-lived. However its bulky fuselage made it suitable as a sleeper aircraft and American Airways and Eastern Air Transport operated the type in both day and night configurations. The sleeper version could carry twelve passengers and a cabin attendant. Condors had retractable undercarriages and could cruise at 167mph over a range of 716 miles.

Fokker F-X and F-XA

These United States-built aircraft were very similar to the Fokker F.VIIb-3m but larger, heavier and more powerful. They had accommodation for twelve passengers. About 60 were built from 1927 and they were quite widely used but on the last day of March 1931 the wing of an F-XA failed in flight. One of the passengers killed was a famous football coach and the accident was almost treated as a national disaster. As a result of the accident passenger carriage in wooden or wooden-winged aircraft was banned.

The cabin of a fourteen-passenger Ford 5-AT Tri-Motor.

Ford Tri-Motor

Regarded as one of the most important aircraft in the history of air transport the Ford Tri-Motor first flew in June 1926. It was a three-engined high-wing monoplane of all-metal construction with corrugated metal skin like the Junkers types in Europe.

Those in airline service were mostly of two versions, the eleven-seat 4-AT and fourteen-seat 5-AT, the latter having 420/450hp engines and a cruising speed of 115mph. Nearly 200 were built up to September 1932 and a few are still flying. The Ford Tri-Motor is probably best remembered for its inauguration of US transcontinental services which were partly by air and partly by train. The Airway Limited train operated overnight between New York and a purpose-built railway station/airport terminal at Port Columbus, Ohio. From there the Fords flew to Waynoka, Oklahoma, with lunch served on board. Trains took over at Waynoka and made the night journey to Clovis, New Mexico, where Fords then took on the passengers for the final stage to Los Angeles. The entire journey took 48 hours. Ford Tri-Motors served numerous airlines in the United States and did much work as passenger and cargo aircraft in Latin America.

Ford 14-A

The final development of the Ford Tri-Motor was the 110ft span 40-passenger 14-A which had two engines buried in the wing with a third mounted above the fuselage. It had a main cabin, smoking room, two lavatories and a galley but it only did taxiing trials and never flew.

CHAPTER NINE
EUROPE - EARLY 1930s

The first of eight Armstrong Whitworth A.W.XV Atalanta monoplanes built for Imperial Airways. It was the first British mainline monoplane. (Flight)

During this period one very specially-designed aeroplane was built for Imperial Airways, the famous Junkers-Ju 52/3m began its long career and Fokker built the last of its three-engined monoplanes.

United Kingdom Armstrong Whitworth A.W.XV Atalanta

In January 1932 Imperial Airways extended its United Kingdom-Central Africa route to Cape Town and required new aircraft for this very difficult operation which encountered very high temperatures and had aerodromes at elevations of up to 6,000ft.

In 1929 the airline had made the decision that all its future aircraft should be equipped with four engines and part of the specification for the new African-route airliner stated that it should have the ability to maintain a height of 9,000ft with any of the four engines stopped while carrying a payload of 3,000lb.

The Atalanta, which first flew in June 1932, was a high-wing monoplane of mixed construction with four 340-375hp air-cooled radial engines. It

Part of the cabin of an Atalanta. The walls were painted yellow in the belief that flies would not land on that colour.

The Junkers-Ju 52/3m was almost certainly the most widely used European transport aeroplane of the 1930s. Part of Lufthansa's large Ju52/3m fleet is seen at Tempelhof Airport.

was reported that the Atalanta could take-off at maximum weight with a run of only 375 yards but this was probably at sea level and in moderate temperatures. Very great care was taken to ensure maximum comfort when flying in the tropics and its cabin walls were painted yellow in the belief that flies would not alight on that colour. The spacious cabin had nine reclining seats arranged in two pairs of forward-facing seats on the port side, each pair with its own table, and forward of these seats was a pair of single seats facing across a table. On the starboard side were two seats facing across a table and forward of that a single seat with table. There were heating and ventilation controls above the windows and there was a lavatory at the rear of the cabin.

Imperial Airways described the eight Atalantas as "the most luxurious and fastest air liners yet produced to travel in the tropics". It was the airline's first mainline monoplane.

In addition to operating the Kisumu-Cape Town services Atalantas worked the routes east of Karachi, two being registered to Imperial Airways' associate Indian Trans-Continental Airways. They continued in operation up to the war when the five survivors were impressed for military service. Atalantas had a cruising speed of 118mph, a maximum

View forward in one of AB Aerotransport's Junkers-Ju 52/3ms.

ABOVE: One of five Fokker F.XVIIIs built for KLM in 1932. This one was named Pelikaan and at Christmas 1933 it carried mail from Amsterdam to Batavia in a flying time of 73.5hr.
BELOW: A KLM Fokker F.XVIII, with thirteen seats, used on European routes after being replaced on the Eastern route by Douglas DC-2s. (KLM)

speed of 156mph and a range of 640 miles.

On some routes seating capacity was raised to seventeen.

Germany
Junkers-Ju 52/3m

One of the best known pre-war European transport aircraft was the much loved Junkers-Ju 52/3m three-engined low-wing monoplane which was built in greater numbers than any other European airliner. Total production has been reported as 4,835 but some sources have quoted over 5,000.

However, this outstanding aeroplane began as a single-engined freighter and the first five were built in this form. This original type first flew in October 1930 and the three-engined type in April 1932.

The Ju 52 followed the configuration and structure of the earlier Junkers transports and was the last to have the well-established corrugated metal skin.

There was normally accommodation for fifteen passengers with six single seats on each side and a triple seat aft. The

ABOVE: The cabin of a Ju 52/3m looking aft

The sole example of the twelve-passenger Fokker F.XX.

cabins were well appointed but the aeroplane was noisy.

Ju 52/3ms were widely used by European airlines, Lufthansa had 78 at the end of 1940, and some twin-float versions served the Norwegian coastal route to the Arctic, Swedish Air Lines and airlines in South America. Lufthansa still has one example for special occasions, there is one in South Africa and three or four are flying in Switzerland. In addition to its great contribution to commercial aviation large numbers, said to be 3,500, served with the Luftwaffe including the invasion of Crete.

The Netherlands

The cabin of the Fokker F.XX Zilvermeeuw (Silver Gull).

Fokker F.XVIII

Five of these aircraft were built in 1932 to operate KLM's Amsterdam-Batavia services. They resembled previous Fokker transports but were larger. They were the last of the flat-sided Fokkers with non-retractable undercarriages.

On the Batavia route there were four sleeper seats and the radio operator and navigator were also accommodated in the passenger cabin.

In June 1935 the Fokkers were replaced on the Batavia route by Douglas DC-2s and the F.XVIIIs became thirteen-passenger European aircraft.

In 1933, Pelikaan (Pelican), the second F.XVIII flew Christmas mail from Amsterdam to Batavia with a flying time of 73hr 34min and the next Christmas Snip (Snipe), the last of the five, carried mail from Amsterdam to Curaçao, a distance of 6,400 miles in 55hr 58min.

Fokker F.XX

In 1933 a very different Fokker airliner was delivered to KLM. This was the F.XX Zilvermeeuw (Silver Gull) which continued the three-engined high-wing monoplane configuration but had lower slung engines, an elliptical-section fuselage, trailing-edge flaps and a retractable undercarriage. There was accommodation for twelve passengers with single seats on either side of a central aisle. The only example was used on European routes.

CHAPTER TEN
SOME EXCEPTIONS

Hannibal, the first of eight Handley Page H.P.42 and H.P.45 biplanes built for Imperial Airways. (Charles E Brown)

Most of the aircraft so far described carried their passengers in what can best be described as a rectangular box with windows. However, there were some exceptions. Flying-boats were providing a higher standard of passenger comfort and there were a few landplanes with greater capacity and superior passenger accommodation.

United Kingdom
Handley Page H.P.42
Hannibal class

Imperial Airways continuing with its four engines policy ordered eight large four-engined airliners for its European and Empire routes.

Initially there were four H.P.42 Hannibals for Empire routes and four H.P.45 Heracles class for Europe and the first to fly, on 14 November 1930, was Hannibal. It went into service first on 11 June 1931 on the London-Paris route but was transferred to Egypt from where the type worked to Kisumu in Kenya and Karachi then in India.

The Hannibal had two passenger cabins - then called saloons - as far as possible from the wing-mounted engines. The cabins were about 8ft wide with the rear cabin having twelve or fourteen seats and the forward cabin six or ten. On the starboard side between the cabins were mail and baggage holds and on the port side a lavatory, bar, steward's seat and cabin-luggage hold. Some of the seats, all in

ABOVE: *The forward cabin of a Handley Page H.P. 42 Hannibal class.*
BELOW: *View aft from the front of the forward saloon of a Heracles class H.P.45.*

Part of the aft saloon of the Heracles class H.P.45.

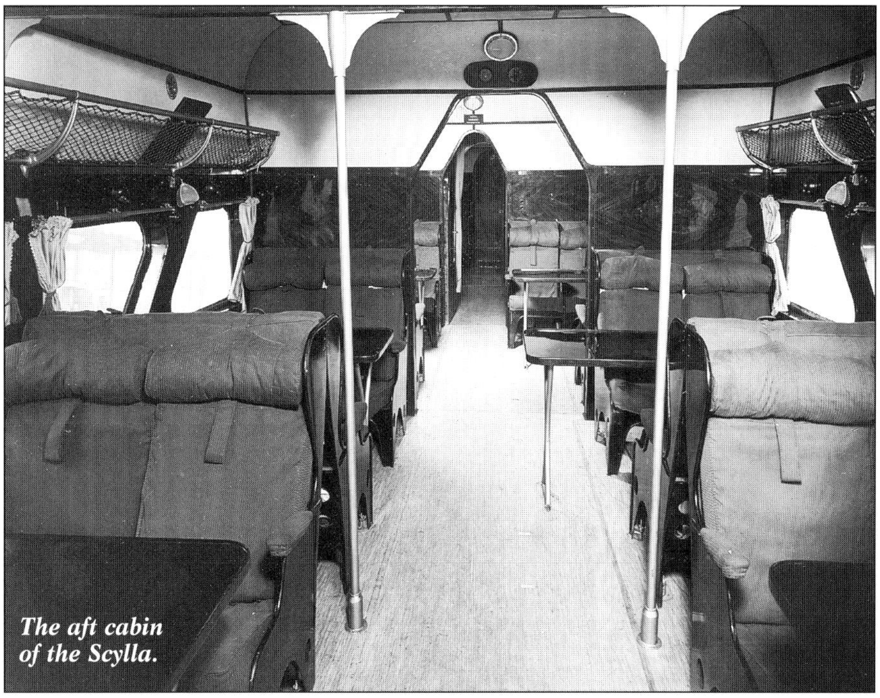

The aft cabin of the Scylla.

The galley of the Scylla. (Weekly Illustrated)

pairs, faced across tables on which full meals were served. The windows were curtained and above them were light-luggage racks.

Two Heracles aircraft were converted to Hannibal standard and one Hannibal was converted to European type.

Handley Page H.P.45 Heracles Class

The H.P.45s and 42s were almost identical apart from interior accommodation. Heracles went into service on the London-Paris route on 11 September 1931 and was at that time probably the most comfortable of all airliners, the two saloons being comparable to Pullman Cars. The rear cabin had eighteen or twenty seats and the forward cabin sixteen or eighteen - two seats in each cabin being optional. All seats were in pairs and chintz-covered. Between the cabins on the starboard side was a baggage hold and to port were two lavatories and the steward's pantry.

The only drawback was the low cruising speed of about 100mph but this allowed the two stewards to serve five-course lunches and seven-course dinners on the 200 miles London-Paris route. These large Handley Page biplanes had very short take-off runs and they landed at a very modest 60mph.

Short L.17 Scylla

In the Spring of 1933 Imperial Airways was short of aircraft to cope with increasing traffic on European routes and tried to buy two more H.P.45s but without success. To overcome the situation the airline asked Short Brothers to build two landplane versions of the Scipio class S.17 Kent flying-boats.

The new aircraft retained the wings, engines and tail units of the flying-boats but had newly designed fuselages and land undercarriages. The two L.17 Scyllas were rapidly

Britain was still building large biplanes as late as 1934. This is the Short L.17 Scylla.

constructed and Scylla entered service on the London-Paris route on 16 May 1934.

The fuselages were nearly 11ft wide and accommodation was divided into three cabins with nineteen seats in the aft cabin and ten seats in both the midship and forward cabins. Three starboard seats in the rear cabin and one in the midship cabin were optional but when fitted gave three-abreast seating. The front seats in the midship cabin and rear seats in the forward cabin were three-abreast and there was an optional seat against the forward bulkhead. All seats on the port side were in pairs. Between the front and midship cabins were two lavatories and a baggage hold on the starboard side and steward's kitchen on the port side. Aft of the passenger accommodation was a mail and freight hold. A disadvantage of the Scyllas was their high-positioned tailplanes and on at least two occasions they were blown over. After one incident, Syrinx was furnished as a mock-up for the Empire Flying-boats then under construction.

Germany
Dornier Do X

On 25 July 1929 Dornier first flew its very large Do X flying-boat which had been designed for transatlantic operation. Its span of 48m (157ft 5.75in) and length of 40.05m

The aft cabin in Syrinx, the second Short Scylla, after being furnished as a mock-up for the Short Empire Flying-boats.

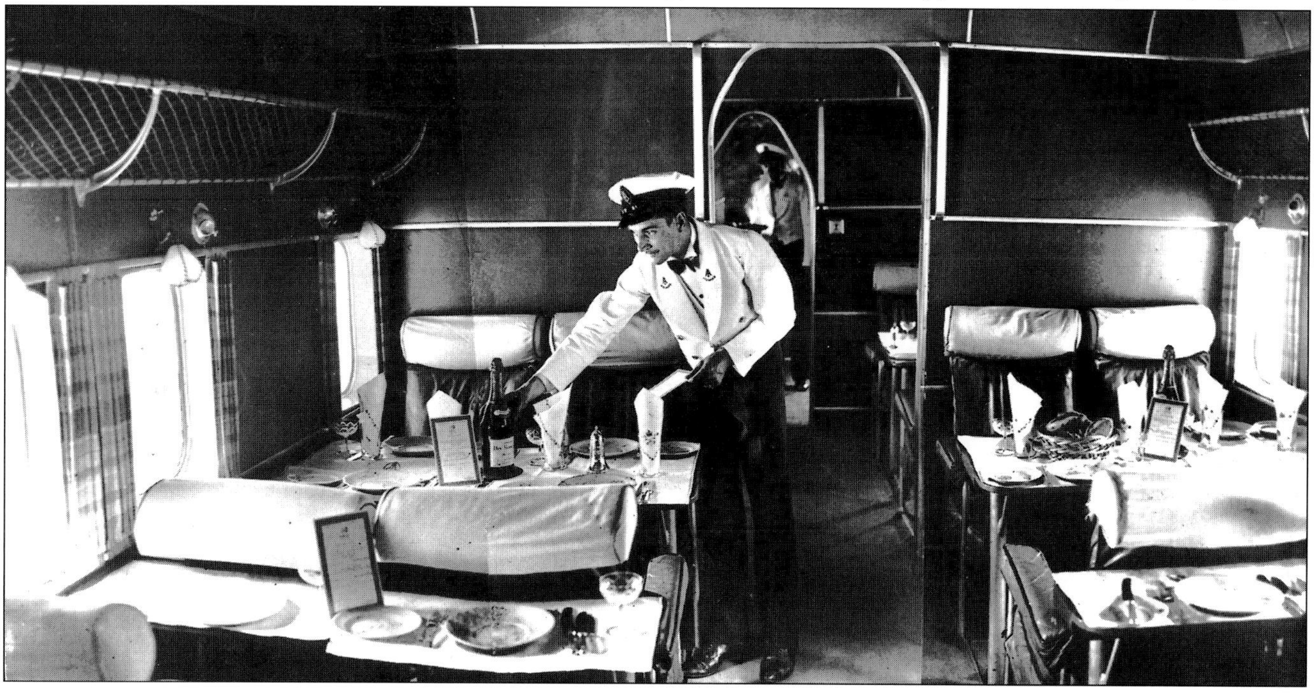

One of the two Fiat-engined Dornier Do X flying-boats built for Italy. The original Do X, then the world's largest aircraft, first flew in July 1929. The Do X made crossings of the South and North Atlantic but never went into service. It once flew with ten crew, 150 passengers and nine stowaways, the largest number of people carried in an aeroplane up to that time. (Dornier)

(131ft 4.75in) made it the largest aeroplane produced by that time. The Do X was designed to carry 66 passengers on long stages and 100 on short stages but on 21 October 1929 it carried 169 people on an hour's flight - this included nine stowaways.

There were twelve engines mounted in tandem pairs above the wing and the hull had three decks, the main deck having the passenger accommodation which was of extremely high standard with rugs on the floor and some seats arranged in groups. The windows were curtained.

The Do X made a protracted flight to New York via South America but never went into service. Two more examples were built for Italy for airline work but this never happened and they passed to the military.

Junkers-G 38

As far back as 1909 Prof Hugo Junkers worked on designs for flying-wing airliners capable of crossing the Atlantic with 100 to 1,000 passengers. The nearest he ever got to producing such aircraft was the G 38 which first flew on 6 November 1929. Its large bat-like wing spanned 44m (144ft 4.25in), had a chord of 10m (32ft 9.5in) and was of 1.7m (5ft 7in) maximum thickness. There were four engines, the structure was all-metal and it had corrugated metal skinning. There was accommodation for 30 passengers.

The G 38 was delivered to Luft Hansa but never went into regular service. However, a much modified G 38b was built and delivered to the airline in September 1931. It had a much deeper fuselage with accommodation on two levels. The first aeroplane was rebuilt to the same standard. The crew's position was above the leading edge of the wing and aft of this was a buffet and an eleven-seat cabin. From the rear of this cabin four steps led down to two cabins with a total of eleven seats. Right aft was a four-seat section. On the wing on each side were three-seat cabins and there

The passenger accommodation in the Dornier Do X was only surpassed by that in the last German Zeppelins. (Dornier)

were two seats in the nose. The whole interior was of a very high standard.

The first aeroplane was lost in an accident in 1936 but the second worked until the war when it was taken for military service. It was finally destroyed in an Allied bombing raid on Athens.

The second Junkers-G 38, delivered to Luft Hansa in September 1931, had three-seat cabins in the wing centre section on each side of the fuselage and a two-seat cabin in the nose. (Lufthansa)

France
Farman F.224

The large four-engined Farman F.224 was exhibited at the 1936 Paris Air Show equipped with 40 seats. Its interior was of high standard and Air France ordered six but rejected them and they were taken over by transport groups of the Armée de l'Air in 1939.

USA
Fokker F-32

In 1929 the Fokker F-32 was built as a 30-passenger aircraft, seven were produced and five were operated by Western Air Express. Unlike previous Fokker transports, the F-32 was four-engined with the engines in tandem pairs suspended from the high-mounted wing. Seats were arranged in pairs on each side of the central aisle and sixteen sleeping berths could be fitted although they were probably never used. A Western Air Express advertisement in 1930 contained the statement "No larger ships, no liners more luxurious exist in America".

ABOVE: View from the G 38's main eleven-seat cabin to the lower-level cabins. (Lufthansa)
BELOW: View forward from the smoking cabin of the Junkers-G 38. (Lufthansa)

View through the G 38's upper cabin to the cockpit. (Lufthansa)

The 40-passenger Farman F.224 ordered by Air France, but never used. It was exhibited at the 1936 Paris Salon.

Part of the interior of the Farman F.224, with double seats on each side.

Dining layout in the Fokker F-32.

View aft through the cabins of the Fokker F-32.

View forward through the cabins of the Fokker F-32.

The 32-passenger Fokker F-32 of 1929 was the first United States four-engined airliner. Western Air Express operated six of the seven built. (Courtesy TWA)

CHAPTER ELEVEN
THE 'MODERN' AIRLINER

The ten-passenger Boeing 247 of 1933 has long been regarded as the first 'Modern' airliner. NC13367, illustrated, was one of the large 247 fleet operated by United Air Lines.

The bans on passenger carriage in wooden aeroplanes and at night or over hazardous terrain in single-engined types had a major influence on future United States designs. From the early 1930s appeared what has been termed the first of the 'modern' airliners, that is multi-engined and of all-metal construction. The first of these was the Boeing 247 and it was soon followed by the great family of Douglas Commercial aircraft. In Europe France and Italy produced similar designs although Savoia Marchetti continued production of a range of three-engined wooden aeroplanes. As mentioned in Chapter Ten in Britain Imperial Airways took delivery of large biplanes as late as 1934.

USA
Boeing 247 and 247D

The first of the 'modern' airliners was the Boeing Model 247 which first flew on 8 February 1933 and entered airline service on 30 March that year. A twin-engined all-metal low-wing monoplane with retractable undercarriage and ten passenger seats the 247 embodied much of the experience gained by Boeing in producing the twin-engined B-9 bomber and the single-engined Model 200 Monomail.

The 247 had a narrow fuselage with five seats on each side and a narrow central aisle which was obstructed by the wing spar passing through the fuselage.

Although an advance on previous types the 247 had poor performance at some of the high elevation airports in the Rocky Mountains. To overcome its shortcomings the re-engined Model 247D was produced and the total number of both types was 73 or 74, with two going to Luft Hansa. Most 247s were brought to 247D standard.

The cabin of the Boeing 247.

A very important aeroplane, the Douglas DC-2 (Douglas)

Boeing was, at that time part of the United Air Lines Group which included Boeing Air Transport, National Air Transport, Pacific Air Transport and United Air Lines and the Group ordered 60 Model 247s — a very large order for that time and one that would lead to the production of the long line of Douglas airliners.

Douglas DC-1 and DC-2

TWA wanted to acquire Boeing 247s to meet the competition from United Air Lines but because of the size of the United order Boeing was unable to supply them. TWA therefore issued a specification to the aircraft industry for a twelve-passenger three-engined all-metal airliner capable of cruising at 146mph and having a range of 1,000 miles. It also had to be able to climb at 1,200ft/min up to a height of 21,000ft.

Douglas met this request but persuaded TWA that this performance could be achieved with a twin-engined type. This was built as the sole twelve-passenger DC-1 which flew on 1 July 1933. The DC-1's performance was so impressive that TWA placed an order for the slightly enlarged fourteen-seat DC-2 of which 198 were built for the United States and a number of overseas airlines including KLM and Swissair. Cruising speed was 190mph at 8,000ft and top speed 210mph. The DC-1 and DC-2, unlike the Boeing 247, had unobstructed floors.

Looking aft through the cabin of a Ceskoslovenská Letecka Spolecnost Douglas DC-2.

Douglas DST and DC-3

American Airlines, successor to American Airways, required a more modern aeroplane to replace its sleeper Curtiss Condors. The fuselage of the DC-2 was not wide enough to accommodate sleeping berths so Douglas designed the DST (Douglas Sleeper Transport) with increased dimensions. This was to prove one of the outstanding stages in airliner development because the dayplane version was the DC-3 of which 10,654 civil and military versions were to be built by Douglas and more than 6,000 under licence, mostly in the Soviet Union as

Flagship Westfield was one of American Airlines' large fleet of Douglas DC-3s. Externally the DST and DC-3 were identical except for the upper berth windows in the DST. The DSTs and pre-war DC-3s, like most other US transport aircraft had their entrance doors on the starboard side. (American Airlines)

PS-84s and Lisunov Li-2s. In British military service this was the Dakota, mostly C-47s. Like the DC-2, the DST/DC-3 was a twin-engined low-wing cantilever monoplane with cellular-structured wing and retractable undercarriage.

The fuselage internal width was 7ft 8in and maximum cabin height 6ft 6in. The DST had fourteen sleeping berths in two tiers. The berths were 6ft 5in long and the upper berths were 2ft 6in wide. Upper berth windows in the DST were the only features which externally distinguished the DST from the DC-3.

So important in air transport history are the DST and DC-3 that it is worth quoting the original Douglas brochure on the types.

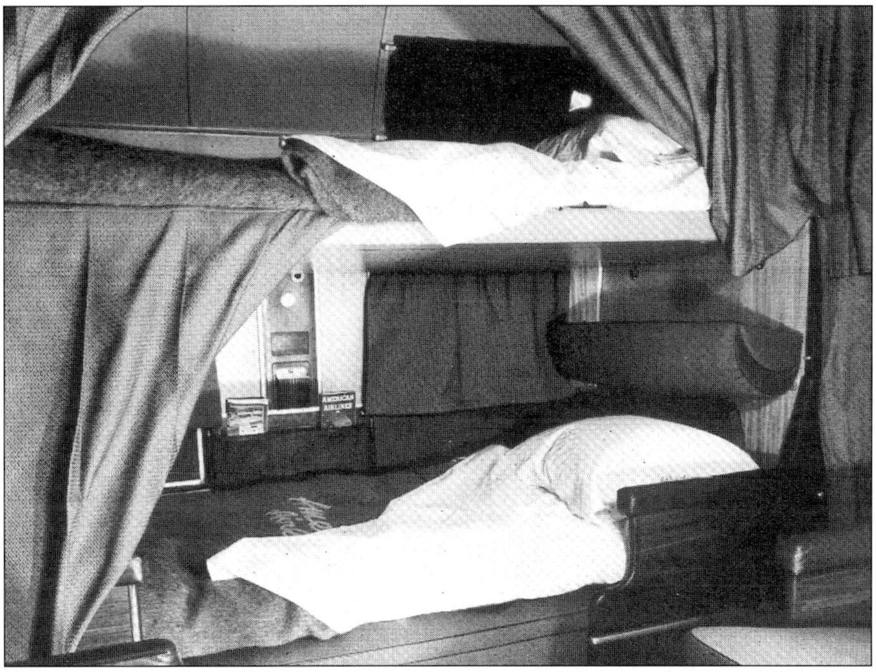

ABOVE: Sleeping berths in an American Airlines' Douglas DST. BELOW: The start of a great series. An American Airlines' Douglas DST (Douglas Sleeper Transport) ready to leave Los Angeles on a night flight to New York.

One of pre-war British Airways' Lockheed Model 10 Electras.

"In the tradition which the DC-2 has established for luxury, security and economical performance, Douglas now presents the DC-3 Day Plane and the DST Sky Sleeper.

The discriminating traveller will find in the saloon of these new Douglas Transports a quiet elegance comparable with that of a palatial ocean liner. Latest developments in the science of soundproofing have been incorporated in the air transports to achieve a reduction in the sound level to approximately 72 decibels at cruising speed.

The air for saloon ventilation, which is admitted through a vent in the nose of the fuselage and circulates through the interior, is as pure as the upper atmosphere of the upper altitudes at which the transports fly. The entire air content is changed every minute. The fresh air is pre-heated by a steam system before entering the saloon and, by manual control, can be maintained at 70° Fahrenheit even when the outside temperature is as low as 20° below zero. The new Douglas Sky Sleeper is America's first air transport designed for night travel and provides sleeping accommodation for 14 passengers. For daytime travel, as many as 28 passengers may be transported in comfort, owing to the generous width of each seat - a full 35 inches wide.

Companion plane to the DST Sky Sleepers the DC-3 Day Plane which is available in two saloon arrangements. In one interior, 21 lounge chairs are provided in two rows with double seats on one side

The cabin of a P.L.L. LOT Lockheed Electra. (LOT)

The cabin of a Lockheed Model 14 Super Electra.

A British Airways Lockheed 14. (British Airways Ltd)

[port] and a row of single seats on the other. The alternate interior offers the luxury and comfort of club car appointments, being equipped with deeply upholstered swivel chairs."

The DST first flew on 17 December 1935 and was introduced as a dayplane by American Airlines between New York and Chicago on 25 June 1936. In August 1936 American Airlines took delivery of its first DC-3 thus releasing DSTs for their night-time operation and on 18 September a sleeper service was begun on a coast to coast route with an eastbound schedule of 16hr and westbound time of 17.75hr. DC-3s mostly in C-47 form played a major role in the Second World War and many examples are still in operation more than 60 years after their first appearance. The DC-3 can safely be claimed as one of the great aeroplanes of all times.

Lockheed 10 Electra

The newly structured Lockheed company built a series of twin-engined all-metal transport aeroplanes. They were smaller than the Boeing 247 and Douglas types but suited thin-traffic routes and they were fast.

The first of the Lockheeds was the Model 10 Electra which had ten passenger seats arranged five each side of the central aisle. There were five large windows each side

ABOVE: View through the aft sixteen-passenger cabin of a Bloch 220.

BELOW: Air France's Bloch 220 Gascogne at Le Bourget with a Farman F.224 in the background.

Air France's Dewoitine D.338 Ville d'Orleans shares the Le Bourget apron with Bloch 220 Provence.

and light-luggage racks extended the length of the cabin on each side.

The first Electra flew on 23 February 1934 and a total of 149 examples were built.

Lockheed 14 Super Electra

In July 1937 Lockheed flew a much refined and larger fourteen-passenger aeroplane. It was of improved streamlined shape and introduced several new features including Fowler flaps to decrease take-off and landing speeds, fully-feathering airscrews and integral fuel tanks.

Lockheed built 112 Super Electras and a further 119 were built in Japan by Kawasaki and Tachikawa.

Electras and Super Electras comprised the fleet of the private British Airways in the last few years before the Second World War and a military development of the Lockheed 14 was the well known Hudson of which nearly 3,000 were built.

An Air France Dewoitine D.338 for European operations. It had twenty-two seats and a bar.

Lockheed 18 Lodestar

The last of the Lockheed twins was the eighteen-passenger Model 18 Lodestar. This was designed to overcome some of the shortcomings of the Model 14 but did not fly until September 1939.

The Lodestar received limited orders from US airlines but twelve were ordered by Pan American Airways for associated airlines and BOAC (British Overseas Airways Corporation) operated several in the Middle East and Africa during the war.

At least one Lodestar was fitted with bench seats for 26 passengers.

Wibault 283.T12 Le Glorieux (The Glorious) was one of Air France's fleet of eighteen aircraft. Two were put into service by Air Union on the Paris — London route in 1933 under the fleet name Voile d'Or (Golden Clipper). (Air France)

Very many Lodestars were used in military service as C-60s and Lockheed built 625. An additional 121 were built by Kawasaki as Ki-56s.

France
Bloch 220

The European type which came closest to the United States 'modern' airliner was the Bloch 220 designed for Air France's main European routes.

The Bloch 220 was a twin-engined all-metal low-wing monoplane with retractable undercarriage and seats for sixteen passengers. It first flew in December 1935 and Air France ordered fourteen.

Passenger accommodation was in a main cabin and a small forward cabin. The comfortable seats had folding tables in their backs, there were large curtained windows and light-luggage racks. Lights and ventilators were above the windows.

Dewoitine D.338

Although the Bloch 220 followed the US twin-engined configuration, Europe continued to build three-engined aircraft both of metal and all-wood construction.

Dewoitine built a series of metal three-engined low-wing monoplanes including 30 D.338s for Air France's European, Far East and South American routes. The first one flew in 1935, the type had a retractable undercarriage and there was accommodation for twenty-two passengers on European routes, fifteen on African and twelve on Far East services, the last of these being six sleeper seats. The seats on European-operated aircraft could be reclined and were arranged as single units on the port side and double seats to starboard. The aisles were carpeted, the windows had curtains and there were full-length light-luggage racks.

Wibault 280.T - 283.T

An earlier French three-engined type was the Wibault 280.T which was exhibited at the 1930 Paris Aero Show.

It was of all-metal construction and had remarkable performance for its low power and it had accommodation for ten passengers. The Wibaults, of which there were several versions, worked Air France's main European routes until 1937 after which they were relegated to secondary routes after the introduction of Bloch 220s. The cabins of the Wibaults were rather austere as is apparent in the photograph.

Interior of a CIDNA Wibault 282.T12 at the 1932 Paris Salon de l'Aviation.

An eighteen-passenger Fiat G.18. The type first flew in March 1935 and was of similar configuration to the Douglas DC-2.

ABOVE: Looking aft in the cabin of a Fiat G.18V. (Avio Linee Italiane)
BELOW: One of Sabena's Savoia Marchetti S.73s. (The Aeroplane)

Italy
Fiat G.18 and G.18V

Italy produced one type of passenger aircraft similar to the Douglas DC-2. This was the eighteen-passenger Fiat G.18 of 1935. Like the DC-2 it was an all-metal twin-engined low-wing monoplane with retractable undercarriage.

Three G.18s entered service with Avio Linee Italiane in 1936 and they were followed in 1937 by six heavier and more powerful G.18Vs. All nine remained in service on airline work up to the war.

Savoia Marchetti S.73

Savoia Marchetti continued to build airliners with watertight wooden wings and in 1934 produced the S.73 with non-retractable

One of Ala Littoria's Savoia Marchetti S.74s. (D M Stroud collection)

undercarriage and accommodation for four passengers in a forward cabin above the wing spars and a fourteen-seat main cabin at a slightly lower level. An unusual feature for that time was the extent of the passenger accommodation, the cabin extending back almost as far as the tailplane.

S.73s were used by Italian airlines, Sabena in Belgium and Czechoslovakia. SABCA in Belgium built seven S.73s to replace Fokker F.VIIb-3ms on the route from Brussels to the Belgian Congo.

Savoia Marchetti S.74

It was not unusual in the 1920s and 1930s for a transport aeroplane to be designed and built for a specific route. One such type was the Savoia Marchetti S.74 designed for Ala Littoria's Rome-Paris trans-Alpine route.

The S.74 was a four-engined high-wing monoplane with wooden wings and non-retractable undercarriage. The cabin was 7.55m (24ft 9.5in) long and had eight reclinable seats on the port side and sixteen on the starboard side.

ABOVE: The Savoia Marchetti S.83, seen here, resembled the S.73 but was slightly smaller. There was accommodation for ten passengers.

The service ceiling was 7,000m (22,965ft), the three-engined ceiling was 5,000m (16,404ft) and oxygen was carried, the leads being attached to the underside of the luggage racks. There was a bar and a lavatory and underfloor baggage holds.

Only three S.74s were built and during the war they passed into military service.

The 24-passenger configuration in one of Ala Littoria's Savoia Marchetti S.74s. (Ala Littoria)

LZ 127 Graf Zeppelin the most successful of all the Zeppelins was the first aircraft to operate transocean passenger air services. (Lufthansa)

The dining room on Graf Zeppelin. (Lufthansa)

The dining room in the Hindenburg.(Popperfoto)

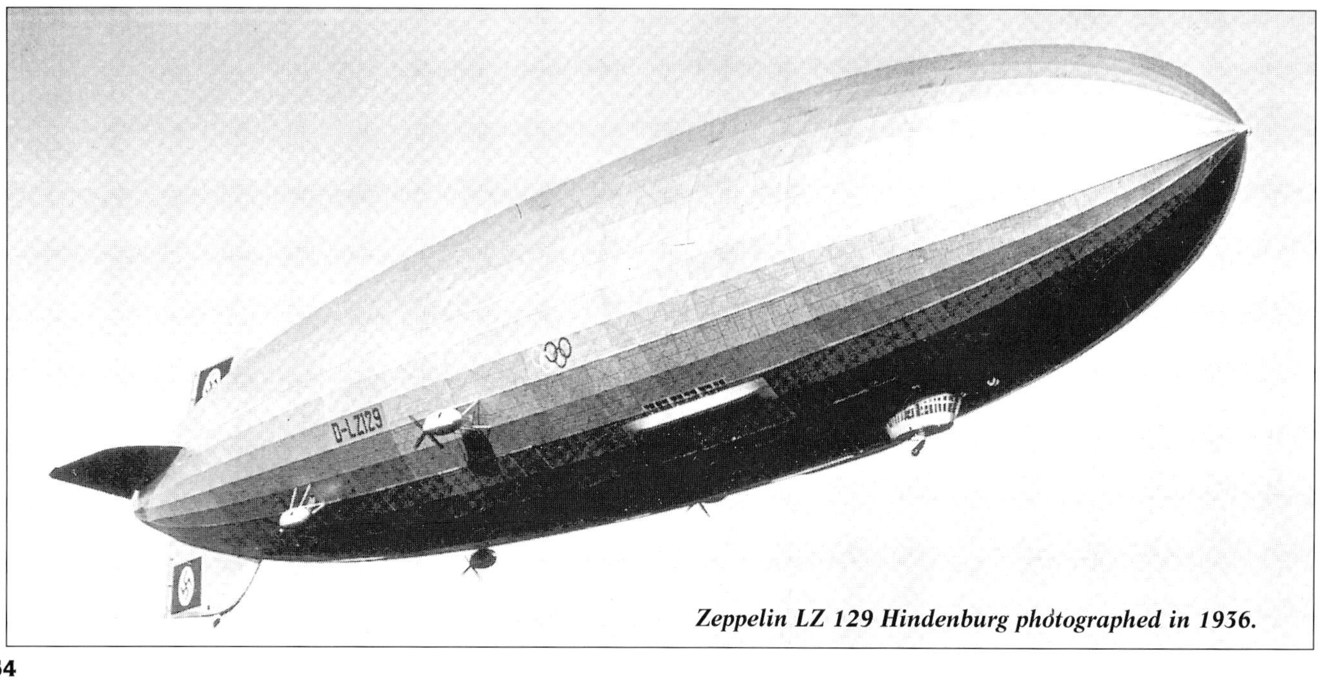

Zeppelin LZ 129 Hindenburg photographed in 1936.

CHAPTER TWELVE
ZEPPELINS - ACHIEVEMENT AND DISASTER

The end of the passenger airship. Hindenburg on fire at Lakehurst, New Jersey, on 6 May 1937 local time. (Popperfoto)

The Locarno Treaty of 1925 removed the Allied restrictions on Germany's building of airships and Luftschiffbau Zeppelin began work on a new rigid airship which was destined to be the best known and most successful of all the Zeppelins. It was followed by a second, larger 'ship and the two new Zeppelins were the first aircraft to operate transocean passenger services.

LZ 127 Graf Zeppelin

LZ 127 was intended for training and research into transocean navigation but it was destined to become the first airship to operate regular long-distance passenger services and the first aircraft to fly passenger services over transocean routes.

Because it was built in the existing shed at Friedrichshafen the LZ 127's dimensions were limited to a length of 236.6m (774ft 3.25in) and maximum diameter of 30.5m (100ft 0.75in). Its gas capacity was 112,000cu m (3,955,240cu ft). There were five 530-550hp Maybach engines, maximum speed was 128km/h (79.6mph) and endurance 118hr.

Graf Zeppelin first flew on 18 September 1928 and normal accommodation was for 36 crew and twenty passengers.

The airship made several outstanding flights, including one round-the-world, before entering regular scheduled service between Friedrichshafen and Recife in Brazil on 20 March 1932. This route was later extended to Rio de Janeiro. It had made the 50th return flight over the South Atlantic route by the end of 1935 and in nearly nine years of operation flew 17,178hr, 1,695,272km (1,053,395 miles) and carried 13,100 passengers.

The control stations and passenger accommodation were in a car beneath the bows, the latter being to a

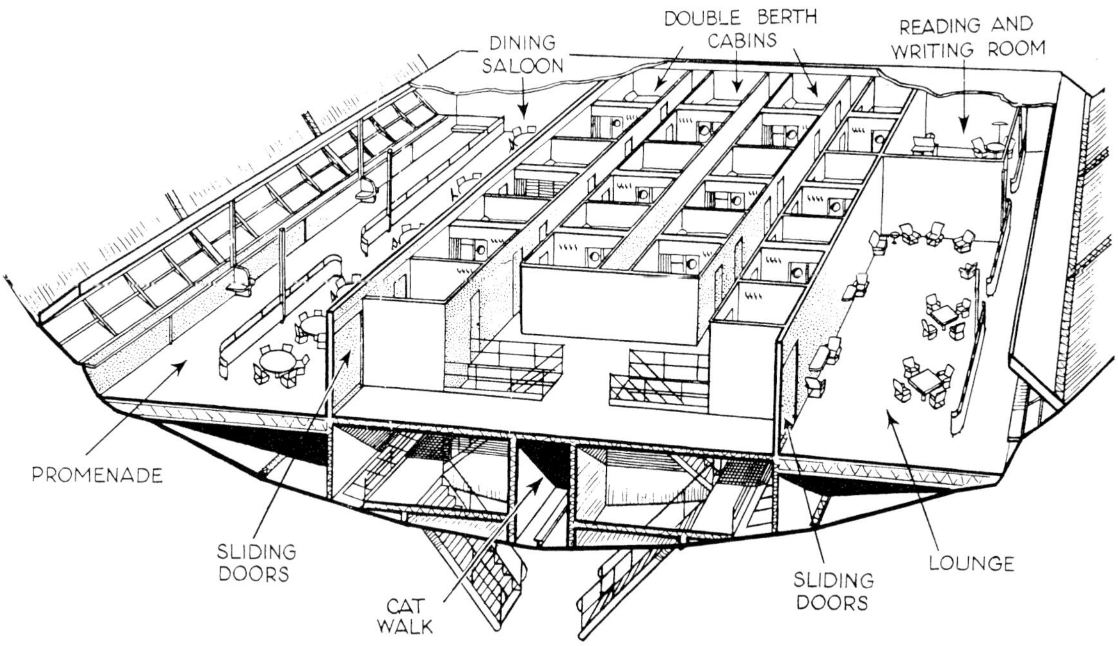

Passenger accommodation in the Hindenburg. (Flight)

very high standard. There was an electric kitchen, a 24-passenger and crew dining room with large windows, ten two-berth cabins arranged five each side of a central corridor, lavatories and wash rooms. Graf Zeppelin was on a return flight from South America at the time of the Hindenburg disaster. It was taken out of service and put on display at Frankfurt-am-Main Airport and was dismantled in 1940.

LZ 129 Hindenburg

The LZ 129 Hindenburg was the first Zeppelin actually designed to carry passengers on transoceanic services and was built in a new shed which removed restrictions on its size. It had a length of 245m (803ft 9.5in), a maximum diameter of 41.2m (132ft 2in) and a gas capacity of 200,000cu m (7,062,900cu ft). There were four 1,050hp Daimler-Benz Diesel engines, maximum speed was 134.98km/h (83.88mph) and endurance 109hr. Normal accommodation was for 40/45 crew and originally 50 passengers but this number was later raised to 72 to meet the demand.

The Hindenburg first flew on 4 March 1936 and at the end of the month made its first voyage to South America, then on 6 May began regular operation of Deutsche Zeppelin-Reederei's service from Germany to New York. In its first season, it flew nearly 250,000km (155,340 miles) and carried 2,057 passengers of which 1,309 were on the New York route. The fastest crossing from Lakehurst to Frankfurt took 42hr 53min.

The Zeppelin was scheduled to make eighteen return services over the North Atlantic during 1937 but on its first service on 6 May, with 97 passengers and crew, after the mooring ropes had been dropped fire broke out and the 'ship was destroyed and 35 passengers and crew were killed or died soon after.

The cause of the fire is still the subject of debate but it brought rigid airship passenger operation to an end although Zeppelin did build the LZ 130 Graf Zeppelin II which flew in 1938.

Hindenburg almost certainly had the finest passenger accommodation of any aircraft. Unlike Graf Zeppelin the passenger accommodation was within the hull and reached by two stairways. On the port side was a dining saloon and on the starboard a large lounge, reading and writing room. The outer sides of these had promenade areas with large outward-sloping windows. The centre of the area was occupied by double-berth cabins.

On the lower deck was a smoking room in addition to a bar, lavatories and shower-bath, officers' mess and kitchen.

In total, Hindenburg made 63 flights and flew 345,303km (214,564 miles).

CHAPTER THIRTEEN
EUROPE - THE LATER 1930s

The first Armstrong Whitworth A.W. 27 Ensign. (Charles E Brown)

In the later part of the 1930s the British aircraft industry began production of its first all-metal landplane airliners although one of the last pre-war types was in fact made of wood. Germany produced some advanced types and Fokker built its last few pre-war airliners.

United Kingdom
Armstrong Whitworth A.W.27 Ensign

In 1934 the British Government announced the Empire Air Mail Programme under which it was proposed, wherever practical, to send all Empire first-class mail by air without surcharge.

To meet the anticipated loads Imperial Airways ordered 28 Short Empire Flying-boats but also

ABOVE: Part of the interior of an A.W.27 Ensign. (The Aeroplane)
BELOW: The prototype de Havilland D.H.95 Flamingo in Guernsey and Jersey Airways colours. (The Aeroplane)

The all-wood de Havilland D.H.91 Frobisher class Albatross. (Flight)

ABOVE: Part of the passenger accommodation in a Frobisher class de Havilland Albatross. All furnishings were in various shades of brown. (BOAC)

A view forward through the cabins of an Albatross. (de Havilland)

decided that a new landplane would be required for the Empire routes and to replace the large biplanes then serving the European routes.

The Armstrong Whitworth A.W.27, called the Ensign class, was produced as the landplane. It was a high-wing monoplane with four Armstrong Siddeley Tiger radial engines and because of the high wing position a very large and troublesome retractable undercarriage. Fourteen were built. There were long delays in production and development testing, the aircraft proved to be underpowered and difficult to fly, and the type did not go into service until late in 1938, on the London — Paris route. It never went into Empire route service before the war.

Empire and European versions were externally identical but there were numerous interior differences. In the European version there were midship and aft cabins each with twelve seats, a forward smoking cabin also with twelve seats and a rear coupé for four passengers. Between the midship and front cabin on the starboard side were two lavatories, a kitchen and a freight and baggage hold. On the port side was the so-called 'Promenade Deck' which was the corridor between the cabins. The coupé had its own lavatory.

The Empire version had midship and aft cabins each with seats for nine passengers or berths for eight. The forward cabin had nine seats or four berths. The Ensigns gave their best service during the war and after being fitted with more powerful Wright Cyclone engines they worked services between Cairo and India.

ABOVE: Junkers-Ju 86 Brocken with Junkers-Jumo heavy-oil engines. A Junkers-G 38 is curving in behind. (Junkers)
BELOW LEFT: The un-furnished interior of a Junkers-Ju 86.
BELOW RIGHT: The Junkers-Ju 86's cabin was only 4ft 9in wide and the seats were slightly staggered and turned inward to provide more room. (Junkers)

The type did not go into post-war service and all the surviving aircraft were broken up.

de Havilland D.H.91 Albatross

One of the most beautiful airliners ever produced, the Albatross had a finely streamlined fuselage, tapered wing and closely cowled engines.

Five of these four-engined aeroplanes were built for Imperial Airways as the Frobisher class and they entered service on European routes in November 1938.

Although it entered service so late, somewhat strangely the Albatross was of all-wood construction. The circular-section fuselage was a moulded sandwich with inner and outer layers of plywood with a filling of balsa.

The new 525hp Gipsy Twelve engines gave the Albatross the remarkable cruising speed of

ABOVE: The prototype Junkers-Ju 90 with Daimler-Benz DB 600 engines. Production aircraft had BMW air-cooled radial engines. (Junkers)
RIGHT: Passenger accommodation in a 40-seat Junkers-Ju 90.
BELOW: An eight-seat compartment in a Junkers-Ju 90.

210mph and a top speed of 225mph. The main undercarriage was fully retractable.

Passenger accommodation was in three cabins, the two front cabins each having four seats a side facing across a table. The aft cabin had the same layout on the starboard side but only one optional forward-facing seat and one double seat on the port side. Normal accommodation was for twenty-two but 23 passengers could be carried. There were two lavatories aft, light-luggage racks on the bulkheads and floodlights above the windows.

Floor widths ranged from 7ft in the aft cabin to 8ft 4in in the front cabin and headroom from 6ft 3in to 6ft 6in. Facing seats were 3ft apart. A plan view of 1938 shows two-tier berths but they were never fitted. There were mail, freight and baggage holds fore and aft.

de Havilland D.H.95 Flamingo

Not long before the war the de Havilland Aircraft Company built its first all-metal transport aeroplane, the Flamingo. The prototype flew in December 1938 and the number built was sixteen.

The Flamingo resembled the Lockheed Electra but had a high-mounted wing and seating for twelve to seventeen passengers. It had two radial air-cooled engines and retractable undercarriage. The prototype

DDL-Danish Air Lines' Focke-Wulf Fw 200 Condor Dania.

was used from May 1939 by Guernsey and Jersey Airways on services to the Channel Islands but the war prevented the airline getting the two it had ordered.

During the war BOAC had seven of the aircraft based in Cairo and in the Middle East.

Germany
Junkers-Ju 86

Germany produced a number of advanced transport aeroplanes during the 1930s. One of these was the Junkers-Ju 86 which was the first type of transport designed to have Diesel or heavy-oil engines. It was a twin-engined low-wing monoplane of all-metal construction, smooth-skinned unlike earlier Junkers airliners and with retractable undercarriage. It had a maximum speed of 375km/h (233mph) and a cruising speed of 360km/h (223.6mph).

Lufthansa is believed to have had fifteen from 1936 and South African Airways had eighteen, a few with Rolls-Royce Kestrel engines.

The Ju 86 was not an outstanding success and the accommodation for ten passengers in the upright-oval-section fuselage was very cramped. The seats were slightly staggered and turned inward to provide leg room.

Junkers-Ju 90

In June 1937 Junkers flew the prototype of its four-engined Ju 90 and Lufthansa had ten or eleven which began operation on the Berlin-Vienna route in 1938. The Ju 90 was a very advanced aeroplane for its time but the war prevented it achieving widescale use.

The type was offered with two interior layouts. In one there were five equal-size cabins each with a facing pair of seats on each side with a table between each group of seats. Smoking was allowed in the two forward cabins.

The other layout had a main cabin with twenty-two or 24 forward-facing seats and a sixteen-seat forward smoking cabin with facing seats. One example is known to have had 40 seats in facing pairs.

View forward in the sixteen-seat rear cabin of a DDL Condor.

This view shows the aerodynamically-clean lines of the Heinkel He 70. (Lufthansa)

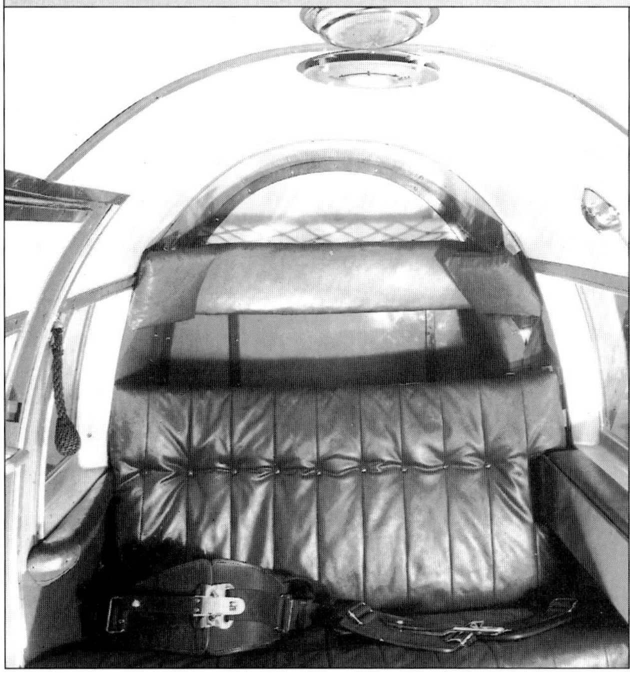

ABOVE: The rear seats in a Heinkel He 70.

ABOVE: The aft cabin of a Heinkel He 111. There was a four-seat smoking cabin forward. (Lufthansa)

Heinkel He 70

When Swissair purchased two of the high-speed Lockheed Orions, it led Luft Hansa to seek faster aircraft and in February 1932 Heinkel was commissioned to design a high-speed four-passenger type. The specification was later changed to achieve 300km/h (186.4mph) and the resultant design was the attractive He 70 single-engined low-wing monoplane with elliptical wing and retractable undercarriage.

The beautifully streamlined metal fuselage contained a small cabin with four seats in facing pairs. Passenger carriage began on 15 June 1934 of Blitz (Lightning) services linking Berlin, Hamburg, Cologne and Frankfurt. They also operated some international European services.

Heinkel He 111

Following successful operation of the He 70 Lufthansa decided that it needed a larger high-speed type with greater reliability and capacity than could be expected from a single-engined type.

The result was the He 111 twin-engined ten-seat low-wing monoplane which bore many of the features of the He 70. There was a main cabin for six passengers and a four-seat smoking cabin forward.

He 111s entered service on German domestic and European international routes and from 1937 were used on some sections of the South Atlantic mail service. Developed from the design was the series of He 111 bombers which became standard equipment of the Luftwaffe.

Focke-Wulf Fw 200 Condor

Almost certainly the best of all the pre-war European transport aircraft was the Focke-Wulf Fw 200 Condor which first flew in July 1937. This was

Leipzig was one of Lufthansa's Heinkel He 111s. (Lufthansa)

an all-metal low-wing monoplane with four engines and retractable undercarriage.

The well-appointed interior had a main cabin with sixteen or seventeen forward-facing seats in double units to port and single units on the starboard side. There was a forward smoking cabin with nine seats. In August 1938 a Condor flew non-stop from Berlin to New York in 24hr 56min and it later flew from Berlin to Tokyo in 46hr 18min with three intermediate refuelling stops. Condors were used by Lufthansa, Danish Air Lines had two as did Syndicato Condor in Brazil. Others ordered by Aero O/Y in Finland and by Japan were not delivered because of the war which prevented the Condor achieving its full potential.

The Netherlands Fokker F.XXII and F.XXXVI

In 1934 and 1935, Fokker produced its last pre-war airliners. These were the F.XXII and F.XXXVI. There were four F.XXIIs and only one F.XXXVI.

They were high-wing four-engined monoplanes with non-retractable undercarriages and seating for twenty-two and 32 respectively. The F.XXII had three six-seat cabins and an aft cabin on the starboard side with two pairs of facing seats.

The F.XXXVI had four eight-seat cabins. KLM used both types and one F.XXII went to ABA — Swedish Air Lines.

The Fokker F.XXXVI Arend (Eagle). It was one of the last pre-war Fokker transports and the biggest of all. Only this example was built.

The Boeing 307 Stratoliner was the first airliner with pressurized cabins. (TWA)

Apart from enjoying the benefits of pressurization passengers on Pan Am's Stratoliners could also watch a film. (Courtesy R E G Davies)

Three-abreast seating in a Pan American Airways' Boeing 307 Stratoliner. (Courtesy of R E G Davies)

CHAPTER FOURTEEN
PRESSURIZATION

Before the Second World War airline operations, with few exceptions, were confined to altitudes not exceeding 10,000ft which meant that it was frequently impossible to avoid unpleasant weather.

The few exceptions to the 10,000ft limit involved the use of oxygen for crew and passengers as on Ala Littoria's trans-Alpine Rome-Paris services.

Boeing produced the first airliner with pressurized cabins but this major development in air transport comfort did not achieve large-scale use until after the war when the United States took the lead in producing outstanding pressurized airliners.

Boeing Model 307 Stratoliner

The Boeing 307 Stratoliner was a most important aeroplane, being the first airliner ever to have pressurized cabins. The type was developed from the B-17C Flying Fortress bomber and combined the wings, power plant and originally the tail surfaces of the bomber with a completely new circular-section fuselage. It originally had accommodation for five crew and 33 passengers. There was for the first time in an airliner a flight engineer's position.

The prototype flew on 31 December 1938 but was lost in an accident which led to design of new tail surfaces.

Three Stratoliners were built for Pan American Airways, five for TWA and one for a round-the-world flight by Howard Hughes.

The fuselage internal width was 134in, cabin length 38ft 2in and cabin height 88in. The cabin area and flight deck were pressurized at 2.51lb/sp in.

The seating was increased from 33 to 38 and accommodation was in four cabins and along the port side. Each cabin had triple facing-seats with tables between them and could be converted to two-berth sleeping compartments curtained off from the single port side seats. There were fore and aft lavatories and wash rooms.

Apart from being the first pressurized airliner the Stratoliner was the first United States-built modern four-engined airliner.

Pan American Airways began operating its Stratoliners, on Latin American routes, on 4 July 1940 and four days later TWA put the type on its New York-Los Angeles route with a schedule of just over 12hr.

Because of the war TWA withdrew its Stratoliners at the end of 1941 and they were bought by the US Army Air Transport Command and operated as C-75s. They were the only type of American transport aircraft capable of crossing the Atlantic and they operated between Washington and Prestwick. By July 1944 they had made about 3,000 accident-free ocean crossings. Pan Am's aircraft maintained Latin American routes under ATC's direction.

In April 1945 TWA's aircraft returned to commercial service on New York-San Francisco and San Francisco-Washington services. Pan Am's Stratoliners then worked for three years between New York and Bermuda.

Stratoliners, with pressurization equipment removed, went to other companies and Aigle Azur operated them in connection with the French-Indochina war. After nearly 50 years one was still in operation with Air Vietiene in Laos.

Average cruising altitude is reported to have been about 10,000ft but its quoted service ceiling was 26,200ft and range 2,390 miles

One of BOAC's Consolidated Liberators used by the airline during the Second World War. (A J Jackson Collection)

During the war BOAC used converted Consolidated Liberator bombers to operate the North Atlantic Return Ferry services and for flights to the USSR. (BOAC)

ABOVE: The cabin of a twelve-passenger Avro York of BOAC. (BOAC)

South African Airways' Avro York Impala at Heathrow. (The Aeroplane)

CHAPTER FIFTEEN
WARTIME AND POST-WAR DEVELOPMENT

One of the Avro Lancastrians used on North Atlantic services by Trans-Canada Air Lines, now Air Canada.

Most British transport aircraft in the early post-Second World War years were adaptations of bombers. France had an airliner which was not completed when France was liberated and Italy produced a development of a wartime type.

Consolidated Liberator

The Liberator was the United States B-24 long-range bomber and several with austere passenger accommodation were made available to BOAC for operation of the North Atlantic Return Ferry service. They also worked other routes including United Kingdom-Moscow and Qantas had some for the Ceylon-Perth route. The photograph opposite shows the cramped conditions.

At the time of its use it was one of the only transports capable of North Atlantic operations. It had a range of 2,850 miles and a cruising speed of 228mph.

Avro York

Britain did produce one wartime transport, the Avro York which first flew on 5 July 1942. The York was a military transport based on the Avro

View aft through the cabin of a Lancastrian. (Avro)

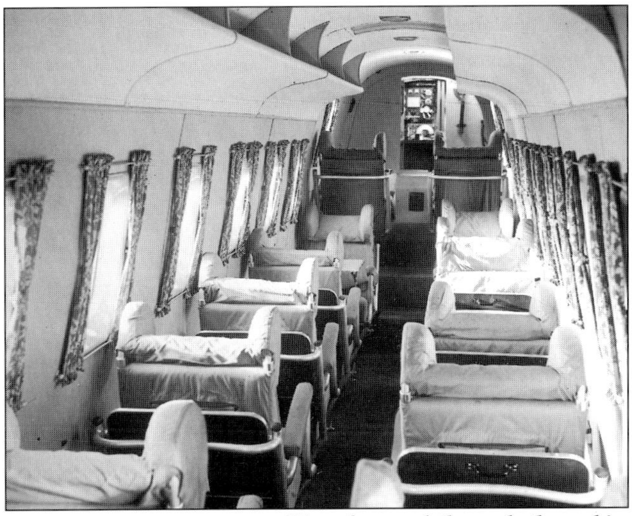

View forward through the cabin of a Lancastrian. (Avro)

BOAC's Handley Page Halton Falkirk. (BOAC)

Lancaster bomber and it used the wing, engines, undercarriage and tail unit of the Lancaster with a new nearly square-cross-section fuselage. The cabin had a length of 54ft, an internal width of 99in and height of 96in. Normal seating was for 24 seated three-abreast with single seats on the port side and double seats to starboard.

Numbers of Yorks were released to BOAC and they proved to be very useful but very noisy aircraft.

Avro Lancastrian

The Lancastrian was a Lancaster bomber with military equipment removed, passenger accommodation provided, a new streamlined nose and new tail cone. It was first produced for a Trans-Canada Air Lines North Atlantic service with ten seats and a range of 4,000 miles. Its maximum speed exceeded 300mph. When the Canadian aircraft were withdrawn in 1947 they had made 1,900 ocean crossings.

Several Lancastrians were made available to BOAC and in conjunction with Qantas they operated United Kingdom-Australia services. The Lancastrian had most unusual accommodation. There were nine inward-facing seats along the port side. Three berths could be made up from these seats and a further three let down to form an upper tier. Passengers could only see out of the starboard side.

The Lancastrian was the first British aeroplane capable of operating services across the South Atlantic and British South American Airways operated eighteen on services to South America. They had more orthodox accommodation for thirteen with single forward-facing seats on each side of the cabin and two front seats facing aft. There were also two forward-facing seats at a higher level over the spar.

Star Dust, one of BSAA's fleet, was lost in the Andes in 1947 and the wreckage was only found in January 2000.

Handley Page Halton

Another bomber conversion was the Handley Page Halton of 1946 which was a passenger adaptation of the Halifax. BOAC had twelve Haltons and

ABOVE: The ten-passenger cabin of a BOAC Halton. (BOAC)

The ninth SNCASE SE.161 Languedoc in its original form with near-circular fins and rudders.

they had ten-seat cabins with four seats on the starboard side and six on the port side. Unusual for a British aeroplane was the entrance door in the starboard side. It was this feature which reduced seating on that side. Haltons operated United Kingdom-Karachi services but were mainly used on routes to West Africa.

SNCASE SE.161 Languedoc

The Languedoc resulted from a 1936 requirement for Air Afrique's short-haul operations. To meet the requirement the 33-passenger Bloch 161 was built and Air France placed an order but it was not until May 1946 that the type was introduced, by which time it had become the SE.161 Languedoc. Sud-Est had delayed production as long as possible because of the German occupation.

From 1946 the Languedoc rapidly became Air France's main European aircraft.

Languedocs appear to have had 24 seats on entering service but in 1951 some were given 44 seats to improve economics and these were used on services to North Africa. Five were also operated by Polskie Linie Lotnicze (LOT) and one was used by Air Liban.

TOP: A Languedoc cabin with 33 seats. Main and forward cabins not separated by a bulkhead. BOTTOM: A Languedoc with 44 seats.

Douglas Dakota

After the war very large numbers of the military versions of the DC-3 became available to the world's airlines and most airlines worldwide acquired fleets of these

A United States Air Force Douglas C-47. Very many became surplus and served the world's airlines for several years. Some are still flying. (USAF)

ABOVE: Early Douglas C-47 conversion for BOAC and BEA. (BEA)
BOTTOM: A British European Airways Douglas C-47 after conversion to 32-seat Pionair class. (BEA)

aircraft, mostly the C-47 version. They were reliable and economic and numerous conversions for civil use were made with 21 seats as pre-war, 28 seats for Tourist-Class operation and as many as 32 in high-density four-abreast layout.

BEA (British European Airways) converted many to Pionair class with 32 seats, two crew and airstairs.

Fiat G.212

In January 1947 Fiat flew the first of its G.212 three-engined low-wing monoplanes and two civil versions were produced mainly for Italian airlines.

The G.212 CP Monterosa was a 34-passenger type with pairs of seats on the starboard side and single seats on the port side. The other version was the G.212 TP Aviocargo Monviso freighter.

Avio Linee Italiane had six G.212 CPs and the Egyptian airline SAIDE had three.

Savoia Marchetti S.M.95

The last Savoia Marchetti airliner broke with the company's three-engined tradition and was a four-engined low-wing monoplane

Avio Linee Italiane's Fiat G.212 Aeropullman Monterosa I-ESTE. (Fiat)

and the biggest of the whole Savoia series.

Normal accommodation was for twenty passengers but LATI's examples used between Rome and Venezuela had 26 seats and it is believed that SAIDE's aircraft had 38 seats.

It was an S.M.95 which opened Alitalia's Rome-London service in April 1948.

For high altitudes there were individual passenger oxygen outlets.

ABOVE: Cabin of a Fiat G.212 CP Aeropullman Monterosa. (Fiat)

Alitalia's Savoia Marchetti S.M.95 Sabastiano Caboto at Northolt on 3 April 1948 after operating the airline's first Rome — London service. (BEA)

TOP: A Douglas DC-4 of AB Aerotransport (Swedish Air Lines)

TOP RIGHT: A Japan Air Lines' Douglas DC-6B. (Douglas)

ABOVE LEFT: View forward in the cabin of an Iberia Douglas DC-4.

CENTRE RIGHT: Looking aft through a Douglas DC-6 (Douglas)

BELOW LEFT: The cabin of an Air France Tourist Class DC-4. Some had 70 seats. (G Boisgontier)

BELOW RIGHT: View aft in a DC-4 converted from a C-54 by Fokker for the Portuguese airline TAP. (Fokker)

CHAPTER SIXTEEN
UNITED STATES AND CANADA LEAD

Competition between airlines operating United States' transcontinental services, competition between aircraft manufacturers and the desire to achieve non-stop North Atlantic services in both directions led to a remarkable series of high-performance piston-engined airliners, in particular the Douglas DC-4 to DC-7 series and Lockheed's Constellation, Super Constellation and Starliner. Boeing produced the luxurious Stratocruiser and to replace DC-3s on short-haul routes Convair built its fine series of Convair-Liners and Martin produced the not so successful 2-0-2 and 4-0-4. The Douglas and Lockheed types proved to be the ultimate piston-engined types up to the introduction of turbojet and propeller-turbine types of the 1950s.

Douglas DC-4

In the mid-1930s American Airlines, Eastern Air Lines, Pan American Airways, TWA and United Air Lines issued a specification for a large long-range airliner and each paid Douglas $100,000 towards the cost of developing the large four-engined DC-4. The DC-4, the first large airliner to have a nosewheel-undercarriage first flew on 7 June 1938 and had limited service with United Air Lines but proved unsatisfactory and no order was placed for this aeroplane which could carry 60 passengers over a 1,000 miles stage.

At the beginning of 1940 the same airlines with the exception of TWA placed orders for 61 of a

Pan American World Airways' Douglas DC-7B Clipper Jupiter Rex. (Pan American World Airways)

redesigned DC-4 and the original aeroplane was redesignated DC-4E.

The new DC-4 first flew on 14 February 1942 by which time the USA was deeply involved in the Second World War and the DC-4s were delivered to the US Army as the C-54 Skymaster and the Navy as the R5D.

At the end of the war many of these aircraft became available to the world's airlines and a further 79 were built as civil aircraft. In June 1946 American Airlines was the first airline to introduce the type. 1,315 were built.

The DC-4 was an all-metal four-engined low-wing monoplane with egg-shaped fuselage cross-section having a cabin length of 55ft 5in and an upper fuselage internal width of 118.5in. In First-Class configuration it had accommodation for 44 passengers, four-abreast. In Tourist-Class there were 66 seats, five-abreast. The DC-4 played a major role in the introduction of Coach-Class in the USA and high-density seating was for 80. Some of the European-operated aircraft had 70 seats. The DC-4 was not pressurized but it was a very fine aeroplane and destined to lead to a family of outstanding airliners.

Douglas DC-6

Lockheed had produced its pressurized Constellation which was faster than the DC-4. It had entered service in February 1946 with TWA and the DC-4 operators required a pressurized development of the DC-4 for their transcontinental routes.

The type produced to meet the requirement was the DC-6 which had the same fuselage but was 7ft longer and had more powerful engines. It could accommodate 50 to 86 passengers and was first put into service by United Air Lines on 27 April 1947.

The DC-6 was followed in 1951 by the DC-6B and the total of DC-6 series aircraft was 704.

Douglas DC-6B

Further stretching of the series

Last of the Douglas piston-engined airliners, the cabin of a DC-7C Seven Seas. (Pan American World Airways)

A Lockheed Constellation of TWA.

came with the DC-6A freighter for Slick Airways and this led to the DC-6B passenger version which was 6ft longer than the DC-6 and with slightly increased power.

The DC-6B first flew in February 1951 and had accommodation for 54 First-Class or 82 Tourist-Class passengers. There could be high-density seating for 102. The type entered service in April 1951 with Western Airlines.

It was to prove the most successful of the family of Douglas four-piston-engined airliners and it has the reputation of being the most economic airliner of its time.

Douglas DC-7

The DC-7 was produced to meet competition from TWA's Lockheed Super Constellation on the US transcontinental routes. It was a stretched version of the DC-6B but with Wright Turbo Compound engines and it was sold exclusively to US trunk route carriers and was put into service in November 1953 by American Airlines.

From the DC-7, the DC-7B with increased fuel capacity for overwater flights was developed and Pan American World Airways put the type into service over the North Atlantic in May 1955. Accommodation in the DC-7B ranged from 69 First-Class to 99 in high-density.

Douglas DC-7C Seven Seas

The DC-7C was the last development of the Douglas range of piston-engined airliners. It retained the same fuselage cross-section but was 2ft 3in longer than the DC-6B. All the previous Douglas four-engined types had a span of 117ft 6in but because of the high noise level of the Turbo Compound engines the DC-7C had a 10ft increase of centre-section span to move the engines a further 5ft from the fuselage.

The DC-7C had a range of 4,000 miles and was capable of non-stop North Atlantic crossings in both directions. It first flew at the end of 1955 and Pan American World Airways introduced the type on New York-London services on 1 June 1956. It had 84 seats

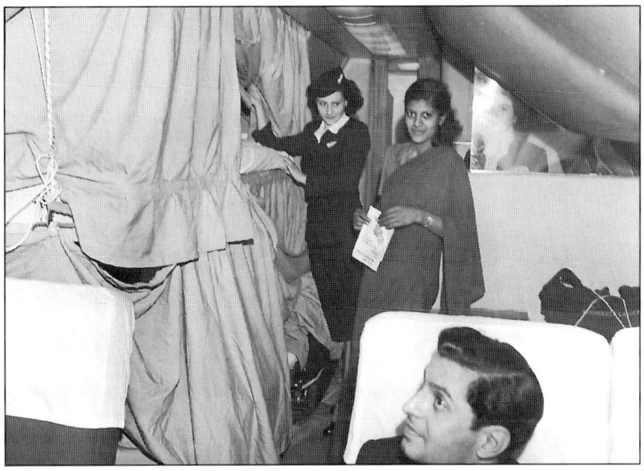

Sleeping berths made-up in Air-India International's Lockheed Constellation Rajput Princess in May 1948. (Flight)

In a Lockheed Constellation of Air-India International. The bulkhead was later removed

A Lockheed L-1049 Super Constellation of Pakistan International Airlines. (Lockheed)

Part of the accommodation in a Lockheed L-1049 Super Constellation.

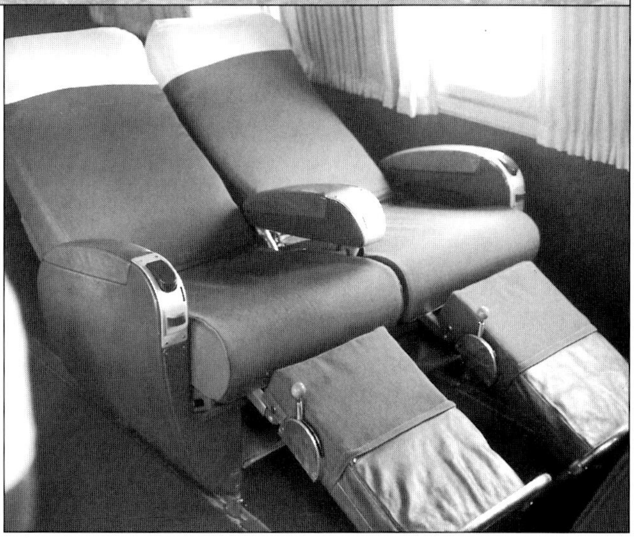

Sleeper seats in a Super Constellation.

and cruised at over 300mph. This was the end of the Douglas line which began with the DC-4 and the turbine-powered aeroplane was to appear on Atlantic routes in 1957. 338 DC-7 series aircraft were built.

Lockheed Constellation

Following discussions in 1939 TWA ordered nine Lockheed Constellations in 1940 and the first aircraft flew in January 1943. As was the case with the DC-4 the war intervened and the type, as the C-69, was first operated by

An Air France Lockheed L-1649A Starliner.

ABOVE: A Boeing Stratocruiser of American Overseas Airlines.
TOP RIGHT: The pantry in a Pan American Boeing Stratocruiser. (Pan American World Airways)
MIDDLE RIGHT: The downstairs lounge of a Stratocruiser, looking forward. (Boeing)
BOTTOM RIGHT: Main Deck of a Pan American World Airways' Boeing Stratocruiser.(Pan American World Airways)

the USAAF from mid-1944. TWA was the first commercial operator beginning services in February 1946.

The Constellation was superior to the Douglas DC-4 in that it was pressurized and considerably faster. Its interior width of 126.8in was slightly greater than that of the DC-4.

Numerous versions of the Constellation were produced, they found worldwide favour and 233 were built. Seating ranged from 44 to 81.

Lockheed L-1049 Super Constellation

In 1950, Lockheed flew the first L-1049 Super Constellation. This type was 18ft 5in longer than the Constellation and carried from 66 to 102 passengers but it was slower than the DC-6B. Further development took place and the L-1049G of 1954, with 63-99 seats, had 3,400hp Wright Turbo Compound engines and was capable of US transcontinental non-stop flights. The final version was the L-1049H freighter and altogether 579 Super Constellations were built.

The Turbo Compound engines were extremely noisy and caused considerable vibration.

Lockheed L-1649A Starliner

What has been described as 'the ultimate in piston-engined transport aircraft development' was the Lockheed L-1649A Starliner which was Lockheed's answer to the Douglas DC-7C.

Lockheed stretched the fuselage of the Super Constellation by 2ft 7in but gave the aircraft a completely new 150ft span wing which put the noisy Turbo Compounds

Martin 4-0-4 City of Los Angeles of California Central Airlines. (R C Steele)

further from the fuselage. The L-1649A could carry 62 to 99 passengers and had a range of nearly 5,000 miles.

The type was introduced by TWA in May 1957. It could cross the North Atlantic non-stop in both directions and some were used on the Polar route to Tokyo flying non-stop between Paris and Anchorage.

But impressive as its performance was the L-1649A appeared too close to the introduction of gas-turbine-powered types and only 44 were built.

Boeing Stratocruiser

Boeing did not enter the Douglas — Lockheed competition with a family of airliners but instead built the unique and, originally, luxurious Model 377 Stratocruiser.

This aeroplane was a civil development of the B-29 and B-50 bombers and the C-97 military transport. It was a mid-wing four-engined monoplane with full pressurization. Its main deck was 78ft 6in long and the upper fuselage interior diameter was 124.5in. Its 55 seats were in main, aft and forward cabins, the last two being separated by dressing rooms.

One of the Stratocruiser's main attractions was a lower deck lounge and bar reached by a spiral staircase from the main deck.

Stratocruisers could operate non-stop eastbound North Atlantic services but had to make refuelling stops on westbound flights. Only 55 Stratocruisers were built and they were introduced on 1 April 1949 by Pan American World Airways. In Tourist-Class configuration, they could accommodate 86-89 passengers and in high-density layout 112.

So spacious was the aeroplane that a steward could walk in front of an aisle seat table to serve a window-seat passenger.

ABOVE: The cabin of a Martin 4-0-4 with full load. (Glenn L Martin Co)

Convair-Liners

The Convair-Liner CV-240 was produced in 1947 to meet an American Airlines' specification for a Douglas DC-3 replacement and first flew in March 1947. It was a pressurized 40-passenger twin-engined low-wing monoplane and quickly found favour as a short-to-medium-haul type and was developed into the larger CV-340 with 44-56 seats. It also had airstairs to the forward door.

Then came the further improved and quieter CV-440 Metropolitan of 1955.

An Alitalia Convair CV-340 Convair-Liner. The airstairs make an interesting comparison with some of the pictures in the early chapters in the book. (Consolidated Vultee)

These Convair-Liners were exceptionally fine aeroplanes providing similar high standard accommodation to that of the trunk route aircraft. Total production was 1,076 and more than 200 were later fitted with propeller-turbines.

Martin 2-0-2 and 4-0-4

Martin produced an aeroplane as a DC-3 replacement. This was the 2-0-2 of 1947, but it was unpressurized and following a structural failure was withdrawn from service. It was followed by the improved and pressurized 4-0-4 which entered service in October 1951 with TWA.

Only 144 Martin 2-0-2s and 4-0-4s were built and the type did not have the success of the Convair-Liners which they resembled.

ABOVE RIGHT: View forward through a Convair CV-340.

BELOW RIGHT: View aft through a Convair CV-340.

One of Trans-Canada Air Lines' Canadair Four North Star Skyliners. (Air Canada)

Canada has a long history of building transport aeroplanes and best known are probably the de Havilland Canada series of short-take-off and landing types such as the Beaver, Otter and Twin Otter and more recently the Canadair series of Regional Jets.

But in the early post-war years Canadair produced a most successful four-engined airliner which served its own airline and air force and was a major type in BOAC's fleet. This was the Canadair Four of which about 170 were built.

Canadair Four

The Canadair Four, which first flew in July 1946, was a Douglas DC-4 powered by four Rolls-Royce Merlin liquid-cooled engines instead of the normal Pratt & Whitney air-cooled radial engines. It was produced to meet a Trans-Canada Air Lines (now Air Canada) specification for a long-haul airliner. There were several versions and the first of them, the DC-4M-1, was unpressurized but TCA took six of them and introduced them on 1 May 1948 on routes serving Calgary and Vancouver. On 1 June 1948 they began Canadian transcontinental service — they worked under the class name North Star.

The fuselage of the Canadair Four had the same cross-section as the Douglas aircraft and the second version, the DC-4M-2, was based on the DC-6 and pressurized. TCA had nineteen of these plus two of a version designated C-4-1. Canadian Pacific Air Lines had four C-4-1s and began operating them over the Vancouver-Hong Kong route in July 1949. BOAC had twenty-two C-4s which it operated under the class name Argonaut and they remained in service until 1958. East African Airways used five and Aden Airways two.

Main cabin of a Canadair Four Argonaut. (Philip Jarrett)

Some Canadair Fours were used by the Royal Canadian Air Force. Normal First Class seating was for 40 and Tourist-Class 56 with five-abreast in the main cabin and the front of the rear cabin.

A disadvantage of the Canadair Four was the noise level created by the Merlins but TCA developed a cross-over exhaust system which made a marked improvement. This was belatedly adopted by BOAC.

CHAPTER SEVENTEEN
EUROPE RE-EMERGES

A Bristol 170 Wayfarer in Channel Islands Airways' colours over Jersey. (Charles E Brown)

The previous chapter clearly shows how the United States led with first-class airliners after the Second World War. There were numerous reasons for this, one being the large US domestic market, an advantage not shared by Europe. In the case of Britain most airliners were designed and built for one or two airlines namely BOAC and the newly created British European Airways (BEA). It was not until the introduction of gas-turbine-powered types that Britain was to play a significant role. It should also be mentioned that the United Kingdom's aircraft industry wasted much effort, resources and money building two very large aircraft that were never used — the Bristol Brabazon landplane and the Saunders-Roe Princess flying-boat. There were also other unrealistic projects which never got further than the drawing boards or wind-tunnels.

ABOVE: Cabin of the Bristol 170 Wayfarer. (Charles E Brown)
BELOW: Passenger section of a Silver City Airways' Bristol 170 Mk 32 Superfreighter. Up to three cars were carried in the area beyond the bulkhead. (Courtesy John W R Taylor)

United Kingdom
Bristol Type 170 Freighter and Wayfarer

The Bristol 170 was initially a military design but the war ended before it could be built. It was considered that there was a need for a utility passenger and freight aircraft, particularly in underdeveloped areas, and so work went ahead on a civil

British European Airways' Vickers-Armstrongs Viking Type 498 Valiant. (BEA)

aircraft although it also achieved numerous military orders.

The 170 was built in two main versions, the Wayfarer with 36 seats and the Freighter which is best remembered for its work with Silver City Airways on vehicle ferry services between England and France when it carried cars in the main hold and passengers in a small rear cabin.

The Type 170 was a twin-engined high-wing monoplane with non-retractable undercarriage and rectangular-cross-section fuselage with an internal width of 96in and maximum cabin height of 79in. In the Freighter, nose-loading doors gave access to the hold.

ABOVE: The tables in an original BEA configuration Viking. (BEA)
BELOW: View forward through a Viking with original configuration (Vickers-Armstrongs)

The first prototype flew on 2 December 1945 and the 32-passenger Wayfarer was used on services to and from the Channel Islands by Channel Islands Airways for a time from May 1946.

More than 200 Bristol 170s were built, the last version being the Mk 32 Super Freighter which could carry three cars and 23 passengers. One of them was converted to 'Super Wayfarer' with 60 seats for Silver City's London — Paris air-coach service. Silver City's fleet carried about 600,000 cars and more than 1.5 million passengers.

Vickers-Armstrongs Viking

The Vickers Viking was built as an interim airliner, as one of many attempts to replace the DC-3. In order to produce a

British South American Airways' Avro Tudor 4 Star Leopard. (Flight)

new aeroplane in the shortest possible time the Viking was to use many components of the Wellington bomber including the fabric-covered geodetic wing and it retained the tailwheel undercarriage. It was a low-mid-wing monoplane and had the disadvantage, like the Boeing 247, of the spar forming an obstruction in the cabin.

The partly parallel section of the fuselage had an internal width of 95in and maximum cabin height was 79in. The aircraft was unpressurized and originally had twenty-one seats. Later versions had 24-27 seats, and eventually, as the Admiral class with BEA 36 seats. The Vikings began service with BEA on 1 August 1946 and for many years was the airline's main type. Numbers of Vikings were exported and total production was 163. One Viking became the world's first jet transport when it was experimentally fitted with Rolls-Royce Nene turbojets and in this form made the first international flight by a jet-powered transport aeroplane, from London to Paris on 25 July 1948.

Avro Tudor

Much of Britain's hopes for the future of its air transport was based on the Avro Tudor, a long and expensive period with little success and many dashed hopes.

In 1943 it was recognised that an interim North Atlantic aeroplane would be required until the Bristol Brabazon was ready. To cover this Air Ministry issued a specification to Avro in March 1944 for a 72,000lb airliner able to carry a 3,760lb payload over 4,000 miles at 235mph.

The resulting aeroplane was the Tudor 1 based on the Lincoln bomber but with new circular-section pressurized fuselage for twelve passengers. It retained the tailwheel undercarriage. Although four 1,750hp engines and a weight of over 70,000lb was not economic for twelve passengers, the interior was to a high standard.

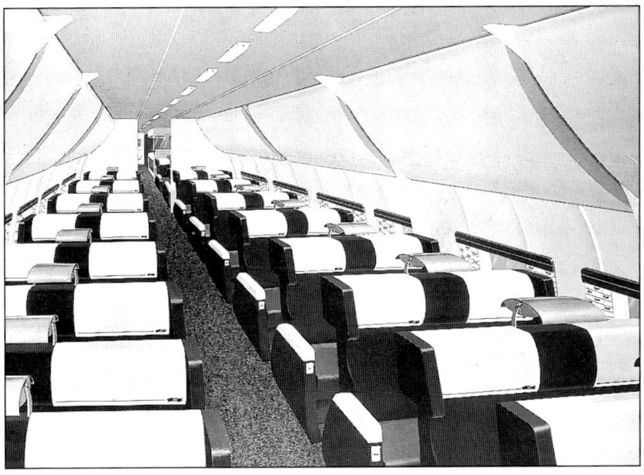
The proposed interior layout for BOAC Avro Tudor 2s, but BOAC never operated any Tudors. (BOAC)

Looking aft through an Avro Tudor 5. This was probably a British South American Airways' aircraft. (Avro)

The beautiful Airspeed A.S. 57 Ambassador. (BEA)

ABOVE: View aft in BEA's Ambassador Sir Richard Grenville. (BEA)
BELOW: View forward in a BEA Elizabethan class Airspeed Ambassador. Seats in the forward part of the cabin were rearward-facing. (Airspeed)

The Tudor 1 first flew on 14 June 1945 and one, named Elizabeth of England, was handed over to BOAC in January 1947 although it was never used.

Much publicity was given to the Tudor 2 and its accommodation as it would be in BOAC service. The Tudor 2 was a lengthened version of the Tudor 1 but like the earlier model it was never used by BOAC.

The only version that had some success as a passenger aeroplane was the 32-passenger Tudor 4 used by British South American Airways but two were lost in unexplained accidents and the type was withdrawn from passenger service although Tudors, unpressurized, continued to work as freighters.

Airspeed A.S. 57 Ambassador

An aeroplane which was much more successful was the beautifully-shaped Airspeed Ambassador built for BEA, but it appeared much too close to the turbine-powered Viscount to achieve a sales success.

Twenty Ambassadors were built for BEA as the Elizabethan class. They were twin-engined all-metal high-wing monoplanes with retractable nosewheel-undercarriage and had pressurized cabins for 40 to 47 passengers. The seats were mostly in twos on the starboard side and threes on the port side and the four rows in the front of the

BOAC's Handley Page Hermes 4 Hero off the Needles, Isle of Wight. (BOAC)

cabin were backward-facing. The seats could be reclined and each seat-back had a folding table. The windows were large and there were full-length light-luggage racks. There were lavatories aft of the cabin and a freight hold forward. The Elizabethan class was put into service in March 1952 only a year before the Viscount.

ABOVE: View forward in a BOAC Hermes 4.
BELOW: View aft in a BOAC Hermes 4. The Speedbird appears on each head-rest. (Flight)

Handley Page Hermes 4

The Hermes 4 was a medium-range airliner for Empire routes and BOAC operated twenty of the twenty-five built. It was a 40-passenger four-engined low-wing monoplane with nosewheel-undercarriage and pressurized fuselage. Internal fuselage diameter was 123in, maximum cabin height 75in and it could accommodate 52 passengers in Tourist-Class or 78 in high density layout. The Hermes 4 entered service with BOAC in August 1950 and after being withdrawn by the airline the type served a number of other operators including British charter companies and Bahamas Airways.

France
SNCASO SO.30P Bretagne

France began development of a 23-seat pressurized transport as early

SNCASO SO.30P Bretagnes of COSARA (Société des Transports Aériens de l'Extreme Orient) in Indo-China.

as 1941. This was the SO.30N Bellatrix but the German Armistice Commission refused permission for it to fly when it was completed at the end of 1942, and it only flew in February 1945.

Based on the SO.30N was the slightly larger SO.30P Bretagne which first flew in December 1947. This was an all-metal twin-engined monoplane with retractable nosewheel-undercarriage and pressurized

The rear section of the cabin in an SNCASO SO.30P. (Rapho)

A view aft through the cabin of an Armagnac. (Scott Henderson collection)

An SNCASE SE.2010 Armagnac in the silver and green livery of Transports Aérienes Intercontinentaux.

An Air France Breguet 763 Provence at Le Bourget in March 1953.

accommodation for 30 or 37 passengers. There were 45 production SO.30Ps but only a small number went into airline service, eight with Air Algérie and twelve with Air Atlas-Air Maroc and its successor Air Maroc. Some of Air Algérie's SO.30Ps had a 160kg (352lb) thrust Turbomeca Palas turbojet beneath each wing to improve take-off and increase the aircraft weight. Two SO.30s were flown experimentally with turbojet power.

SNCASE SE.2010 Armagnac

The Armagnac was a large four-engined monoplane designed to carry 64 passengers across the North Atlantic or up to 107 passengers on short-stage routes. It first flew on 2 April 1949 and production of fifteen began, with the last to have propeller-turbines.

The very deep fuselage was to have had three tiers of sleeping berths but by the time the aircraft was ready sleeping accommodation had gone out of favour and this meant that the Armagnac had a lot of unusable space.

TAI (Transports Aériens Intercontinentaux) took four but returned them as uneconomic. However a new company, SAGETA, used all seven on a supply route between Toulouse and Saïgon and the Armagnacs carried 6,373 passengers over the route, 136,022kg (299,876lb) of cargo and baggage and 29,429kg (69,879lb) of mail. They proved very reliable.

Breguet 763 Deux-Ponts

The Deux-Ponts was a very unusual aeroplane in that it was a true double-deck type with accommodation originally for 59 Tourist-Class passengers on the upper deck and 48 Second-Class passengers or cargo on the lower deck. The fuselage was 5m (16ft 5in) deep and stairways fore and aft connected the two decks which were unpressurized.

Lower deck of a Breguet 763. (Breguet)

Upper deck of a Breguet 763. (Breguet)

Saab 90A-2 Scandia Gardar Viking of SAS.

The rear lower fuselage consisted of four sideways-opening doors. With these large doors open and the aft stairway raised, there was 3m (more than 9ft) clearance for loading large objects into the lower deck.

The Deux-Ponts was a mid-wing monoplane with four engines, retractable nosewheel-undercarriage and twin fins and rudders. Air France operated twelve under the class name Provence and began their operation, between France and North Africa in March 1952.

After their replacement by more modern aircraft a number of the big Breguets were modified to become Universal freighters.

Sweden
Saab Scandia

A newcomer to the design and construction of transport aircraft was Svenska Aeroplan AB (Saab) in Sweden, which flew its prototype twin-engined 24-32-passenger low-wing monoplane Scandia for short-to medium-stage routes in November 1946.

The Scandia was a very nice aeroplane in which to fly but orders were slow and only eighteen were built, six by Fokker in the Netherlands.

Accommodation was later increased to 36. SAS introduced the Scandia on Scandinavian domestic routes in November 1950 and had eight, but Aerovias Brasil acquired the first four production aircraft and eventually all Scandias were taken by the Brazilian airline VASP.

Looking aft through the cabin of a Saab Scandia. (Saab)

Passengers on night flights in SAS Scandias were given leaflets explaining that the exhaust flames were quite normal but could not be seen in the day!

CHAPTER EIGHTEEN
SOME LIGHT TRANSPORTS

RMA Lord Shaftsbury was one of BEA's fleet of de Havilland D.H.89 Dragon Rapides. More than 700 of this 1934 type were built. (BEA)

During the 1930s and again after the war the British aircraft industry produced a range of outstanding light transport aircraft which played an important rôle in the development of domestic air services and found favour in many countries particularly in the British Commonwealth.

de Havilland D.H.83 Fox Moth

This remarkable little biplane could carry pilot and four passengers on the 120hp of its de Havilland Gipsy

The de Havilland D.H.83 Fox Moth of 1932 carried pilot and four passengers on 120/130hp. Its success led to a demand for a bigger aeroplane which was built as the D.H. 84 Dragon. (The Aeroplane)

ABOVE: Draco was one of Imperial Airways' fleet of Diana class de Havilland D.H.86s.
LEFT: The ten-to-twelve-seat cabin of an Imperial Airways D.H.86.
BELOW: The cabin of a West Coast Air Services/Isle of Man Air Services' de Havilland D.H.86.
Seats could be added to the aisle but the safety aspect is rather frightening. Seat-belts were not required in the United Kingdom at that time. (Dorondo Mills)

III or 130hp Gipsy Major engine. The Fox Moth embodied mostly Tiger Moth components mated to a new ply-covered wooden fuselage.

The four-seat cabin was between the wings ahead of the pilot's open cockpit.

The type first flew in March 1932 and in June that year Hillman's Airways introduced two on its Maylands-Ramsgate service. So successful were the Fox Moths that the traffic generated soon called for bigger aircraft.

Ninety-eight Fox Moths were built in the United Kingdom plus fifty-three in Canada and two in Australia.

de Havilland D.H.84 Dragon

The Dragon was a six-passenger biplane, with 4ft wide fuselage,

City of Cardiff **was one of several six-passenger de Havilland D.H.84 Dragons used on United Kingdom domestic services by Railway Air Services. (Railway Air Services)**

powered by two 130hp Gipsy Major engines and it was the aeroplane that met Hillman's Airways' requirement for a bigger twin-engined successor to the Fox Moth.

Dragons were acquired by numerous UK domestic airlines and served in many countries including Egypt, India, Australia and Canada.

It cruised at 109mph with an hourly fuel consumption of only about 13 gallons. One hundred and fifteen Dragons were built in the United Kingdom plus 87 in Australia. It was of great importance because it led to the D.H.89 Dragon Rapide.

de Havilland D.H.86 Express Airliner

The D.H.86 was a large wood and fabric four-engined biplane designed for Qantas Empire Airways' operation of the Brisbane-Singapore section of the England-Australia route. It had accommodation for ten-to-twelve passengers and first flew in January 1934.

In addition to the Qantas aircraft Imperial Airways operated a fleet under the class name Diana. They were used on United Kingdom domestic services by Railway Air Services, some served Misr in Egypt and they also worked in New Zealand and South America.

One of Midland and Scottish Air Ferries' two Airspeed A.S.4 Ferry nine-to-ten-passenger Gipsy-powered biplanes. (Dorondo Mills)

A Westland Wessex of Imperial Airways. One operated by Portsmouth Southsea and Isle of Wight Aviation could have aisle seats added like the West Coast Air Services' D.H.86s.

Most had normal seating but the West Coast Air Services' D.H.86 illustrated was unusual in having aisle seats. Sixty-two D.H.86s were built.

de Havilland D.H.89 Dragon Rapide

The Dragon Rapide was the most outstanding and successful of all the pre-war de Havilland light transports. It was a refined version of the Dragon and incorporated some of the features of the D.H.86 including its finely tapered wings, 200hp Gipsy Six engines and trousered undercarriage.

The Dragon Rapide first flew in April 1934 and

The cabin of an Avro Nineteen. The lavatory, forward of the bulkhead, is enclosed by curtains. (Avro)

The passenger cabin and pilot's cockpit in an Airspeed Ferry.

The Avro Nineteen was a post-war civil development of the military Anson. (Avro)

A six-passenger Spartan Cruiser II of British Airways leaving Newtownards Airport, Belfast, for the Isle of Man. (James Lockhart)

more than 700 were built.

This aircraft had short take-off and landing and good economics and did more than any other type in developing United Kingdom domestic air services.

It is a much-loved aeroplane by those who knew it and several are still airworthy.

Airspeed A.S.4 Ferry

This three-engined ten-passenger biplane was built in 1932 for Sir Alan Cobham's National Aviation Day Displays. Two were used and they gave thousands of people their first flight. Over 90,000 in the first season.

The cabin of a Spartan Cruiser II.

The cabin of the de Havilland Dove looking aft.

The very successful de Havilland eight-to-eleven-passenger D.H.104 Dove first flew on 25 September 1945 and 544 were built. This example was one of those delivered to LAN Chile in October 1949. (de Havilland)

The fifteen/seventeen-passenger de Havilland D.H.114 Heron. This is the Series 2 version with retractable undercarriage. (de Havilland)

Because the Ferry had ten seats it was required to have radio but if radio was installed it could only carry nine passengers and radio was not a requirement. It was ruled that Cobham's aircraft could carry ten passengers without the aircraft having radio.

Only four examples were built, the other two going to Midland and Scottish Air Ferries for services radiating from Glasgow.

Westland Wessex

The Westland Wessex of May 1930 was a development of the Westland IV of 1929. It was a three-engined strut-braced high-wing monoplane capable of carrying up to ten passengers.

One Wessex operated the United Kingdom's first railway air service, between Cardiff and Plymouth and the one used on the Portsmouth-Isle of Wight ferry by Portsmouth Southsea and Isle of Wight Aviation had aisle seats like the West Coast Air Services' D.H.86. Only nine were built.

Spartan Cruiser

The Spartan Cruiser of 1932 was a three-engined low-wing monoplane with seats for six passengers. Its short take-off and landing ability made it useful for operation through small unprepared aerodromes and it was used successfully through the Hebrides including Barra where the beach still serves as the airport.

The cabin of a de Havilland Heron. (de Havilland)

Twelve Cruisers were built and there were three Cruiser IIIs with aerodynamically refined fuselages, trousered undercarriages and seats for eight passengers.

Avro Nineteen

After the war Avro produced the seven-to-nine-passenger Avro Nineteen civil version of the famous Anson reconnaissance type and it enabled Railway Air Services to operate non-stop flights between London and Belfast and London and Glasgow. Unlike all the previous aircraft described in this chapter, it had a retractable undercarriage.

The Britten-Norman Trislander with its unusual engine layout. (Brian Bradbury)

de Havilland D.H.104 Dove

After the war de Havilland Aircraft lost no time in getting back to civil aircraft and flew the prototype Dove on 25 September 1945. Unlike previous de Havilland light transports, the Dove was an all-metal low-wing monoplane with retractable undercarriage. It was powered by two 350hp Gipsy Queen engines and normally had accommodation for eight passengers.

Doves were used by numerous airlines and charter companies and two prototypes and 542 production examples were built.

de Havilland D.H.114 Heron

The Heron prototype first flew in May 1950 and it was designed as a simple economic aircraft capable of operating from restricted aerodromes. It was very similar to the Dove but larger and capable of carrying up to seventeen passengers.

Powered by four Gipsy Queen engines the Heron was produced in two main versions, the Heron 1 with non-retractable nosewheel undercarriage and the Heron 2 with fully-retractable undercarriage.

Herons saw widescale airline use and more than 50 were built.

Britten-Norman Trislander

The Trislander entered service in June 1971 with Aurigny Air Services in the Channel Islands and was an unusual aeroplane. Developed from the very successful twin-engined Islander it had increased dimensions and a third engine mounted on the tailplane and its narrow and very confined fuselage could accommodate up to seventeen passengers. There was a separate door to each 'row' of seats.

The engines were 260hp Lycoming air-cooled units and the aircraft, a high-wing monoplane, had a non-retractable nosewheel undercarriage. Both the Islander and the Trislander have extremely short take-off runs which made them ideal for operation in such areas as the Orkney and Shetland Islands.

The none-too-spacious cabin of the Trislander. Each 'seat-row' had its own door. (Loganair)

The first Short S.8 Calcutta. The type was used by Imperial Airways from 1929 on the trans-Mediterranean sectors of the Empire routes. (Short Bros)

The front of the Sea Eagle cabin hinged back for loading newspapers. (Supermarine)

The fifteen-seat cabin of the Short Calcutta City of Alexandria. (Short Bros)

The wooden Supermarine Sea Eagle had a six-seat bow cabin and three Sea Eagles of British Marine Air Navigation flew the first air services to the Channel Islands, Southampton — Guernsey, from September 1923. G-EBGR is seen at Southampton after passing to Imperial Airways.

CHAPTER NINETEEN
FLYING-BOATS

Canopus, the first of the Short S.23 C class Empire Flying-boats, on the Medway at Rochester. (Flight)

During the early decades of air transport the flying-boat played an important rôle in the development of overwater air services and some of them provided the highest standard of passenger comfort. This was made possible because the hulls were generally wider than landplane fuselages because their width was dictated by the laws of hydrodynamics.

ABOVE: The seven-seat Smoking Cabin of the Empire Flying-boats was immediately aft of the mooring compartment. RIGHT: The Promenade Deck of an Empire Flying-boat.

Pan American Airways' Clippers established trans-Atlantic and trans-Pacific services and the Imperial Airways outstanding Short Empire Flying-boats maintained the Empire routes to South Africa and

Short S.17 Kent Scipio class flying-boat Satyrus flying off Alexandria. This type was the first British four-engined aeroplane to enter airline service. (Charles E Brown)

Australia. Before the Second World War most Mediterranean operations were by flying-boats and during the war Empire Flying-boats provided the only transport link between New Zealand and Australia.

Float seaplanes played a lesser rôle but served in the Baltic, the Adriatic and on the rivers of South America.

The provision of airports world-wide and the availability of superb long-range landplanes led to the demise of the flying-boat except for local services such as those in the Caribbean.

United Kingdom
Supermarine Sea Eagle

In September 1923 British Marine Air Navigation began working Southampton-Guernsey services with three Sea Eagle wooden biplane flying-boats each

***ABOVE:** The forward part of the sixteen-passenger Short Scipio cabin. Note the early example of three-abreast seating against the bulkhead. The entrance hatch is visible through the cabin doorway. (Short Bros)*

***BELOW:** Part of the eight-seat Promenade cabin of an Empire Flying-boat. At a higher level through the door is the six-passenger aft cabin.*

with enclosed accommodation for six passengers in the bows.

The small cabin appears to have been rather cramped and the forward section hinged upward to allow loading of goods. Passenger entry appears to have been via steps to the rear of the cabin and required some smart footwork.

ABOVE: BOAC's Short Sunderland Hanwell was one of the military aircraft with refurnished interior that was operated under the class name Hythe.

RIGHT: View through Short Sandringham. (Short Bros via D M Stroud)

BELOW: BOAC operated a number of Plymouth class twenty-two-passenger Short S.25 Sandringham flying-boats from May 1947. Pembroke is seen here taking-off from Southampton Water. (BOAC)

The last of the big British flying-boats to go into service was the Short S.45 Solent. BOAC introduced Solents on United Kingdom — South Africa services in May 1948. The example illustrated was named Sark.

Short S.8 Calcutta

The Short S.8 Calcutta three-engined flying-boat, largely of metal construction, was designed to operate the trans-Mediterranean section of the Imperial Airways' England-India route.

The Calcutta was a large biplane with a cabin measuring 17ft in length, 6ft 6in in width and 6ft 3in in height. It contained fifteen seats with four rows of three (two to starboard and one on the port side) with an additional pair and a single seat on the starboard side. There were full-length light-luggage racks and electric light. There was a buffet cabinet, an oil-burner cooker and a lavatory. There were folding tables and the seat cushions could be used as life jackets.

Imperial Airways had five Calcuttas and two were built for France.

The Calcuttas began trans-Mediterranean operation on 31 March 1929.

Short S.17 Kent (Scipio Class)

The Scipio class Short Kent was an enlarged development of the Calcutta and it went into service across the Mediterranean in May 1932. The Kent had four air-cooled radial engines and accommodation for fifteen passengers in a cabin 14ft long, 8ft 9in wide and 6ft 6in high. Most seats were in facing pairs with tables between them but there was a triple rearward-facing seat against the forward bulkhead. Interior furnishing was to a very high standard. Aft of the cabin were the galley and lavatory and wash room. Only three Kents were built.

The upper deck aft cabin of a Short Solent. (Short Bros & Harland)

Short C class Empire Flying-boat

In December 1934 H M Government announced the Empire Air Mail Programme as 'Beginning in 1937, all letters despatched from the United Kingdom for delivery along the Empire routes would, so far as practicable, be carried by air without surcharge'. Imperial Airways estimated that this would involve carrying 2,000 tons of mail a year and ordered a fleet of 28 Short S.23 Empire Flying-boats to handle this load. The number was later increased to 31 with six going to Qantas Empire Airways.

From 1934 Air France operated a fleet of ten-to-fifteen-passenger Lioré et Olivier H 242 flying-boats on trans-Mediterranean services. Most were in service until the Second World War.

The Empire Flying-boat, which first flew on 4 July 1936, was a four-engined high-wing all-metal monoplane and it provided a very high standard of passenger comfort.

Passenger accommodation was all on the main deck while an upper deck contained the control section and mail holds. The forward passenger cabin was a smoking section for seven day passengers or four at night. Aft of this was the main entrance door, kitchen and two lavatories. Then came a midship cabin for three day or four night passengers. Behind this cabin was the Promenade Cabin with eight seats or four berths to starboard and standing space and higher level windows on the port side. Right aft at a higher level was a cabin with six seats or four berths bringing the total seating to 24 or 16 berths for night travel. All seats could be reclined and normally a midship berth was made up but it is doubtful whether the other berths were used. Most of the interior furnishings were in various shades of green.

In January 1937 the Empire Flying-boats entered regular service and the Empire Air Mail Programme was inaugurated on 29 June 1937 when RMA Centurion left Southampton with 3,500lb of mail for Sudan, East and South Africa.

Two Empire Flying-boats, Caledonia and Cambria, made North Atlantic survey flights during 1937. In 1938 heavier S.30 'boats were added to the fleet. These were equipped for in-flight refuelling and began a weekly experimental mail service from Southampton to Montreal and New York at the beginning of August 1939. There were also two S.33 'boats at a still higher weight.

Air Orient's CAMS 53 Alsace at Naples. In the 1920s and 1930s a number of French airlines operated trans-Mediterranean services with a series of four-to-six-passenger CAMS flying-boats.

The nineteen/twenty-passenger Breguet 530 Saïgon of 1934. Air France operated the only two built on trans-Mediterranean services. The Saïgon's design owed much to the Short Calcutta.

Tasman Empire Airways (TEAL) in New Zealand had two Empire 'boats and during the war they operated the only passenger service of any kind between New Zealand and Australia.

Some of these fine flying-boats remained in service until March 1947.

Short S.25 Sunderland

During the war a number of Sunderland maritime-reconnaissance flying-boats were demilitarized and supplied to BOAC to supplement the Empire Flying-boats. They had fourteen seats. After the war some were given upgraded accommodation, had the bomb bay replaced as an eight-seat lounge and operated as the twenty-two passenger Hythe class.

The eleven-seat Second Class cabin of a Breguet Saïgon. (Air France)

Short S.25 Sandringham

The Sandringham was a Sunderland with faired bows and rear hull and BOAC operated the type as the Plymouth class with accommodation for twenty-two day or sixteen night passengers. The type was also operated by TEAL in New Zealand, Qantas in Australia, DNL in Norway, and some South American airlines.

Short S.45 Solent

The 39-44-passenger Solent was the last of the big Short flying-boats to go into service. Twenty-three were built and they were operated by BOAC, TEAL, Aquila Airways and in Australia by Trans-Oceanic Airways.

Interior layouts varied but records show the New Zealand aircraft as 40-seaters with accommodation on two levels with five lower deck cabins each having a pair of facing seats on the starboard side and single facing seats to port. Between the two front cabins and the aft cabin were men's and women's lavatories and from the forward of the rear cabins a stairway led to the upper deck which had two six-seat cabins with double seats on the port side. A ladder between the women's toilet doors gave access to the crew's area on the upper deck.

The last Solent service, from Southampton to Madeira, by Aquila Airways ended on 30 September 1958 and with it ended the fine history of British flying-boat

The Latécoère 631 was built to meet a 1936 French Air Ministry specification for a 40-passenger trans-Atlantic flying-boat. The first example flew in November 1942, was confiscated by Germany and destroyed in an Allied air raid. The 46-passenger version was built after the war and entered service with Air France in July 1947. (Charles E Brown)

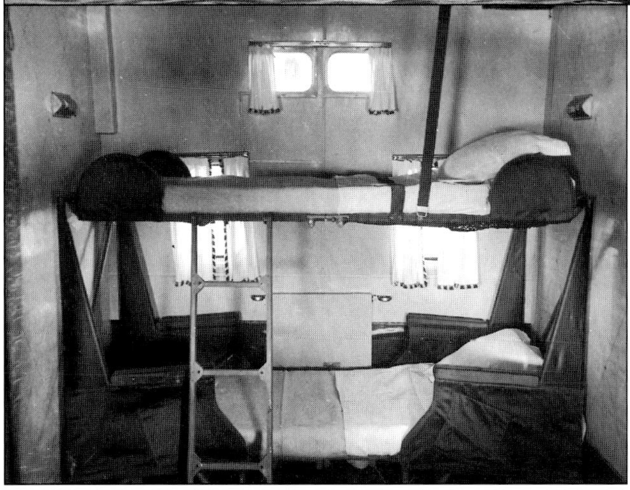

Sleeping berths made-up in a Latécoère 631. (S.C.A.)

The cabin of a Dornier Delphin II.

operation stretching over 35 years.

France
CAMS 53

The CAMS 53 twin-engined wooden biplane flying-boat was widely used on pre-war trans-Mediterranean operations. The first example was introduced on the Marseilles-Algiers route by Aéropostale on 22 October 1928. About 30 were built and there were some larger CAMS 56 and 58 'boats.

The CAMS 53 had accommodation for four passengers in the cabin forward of the wings. Air Union's CAMS 'boats were painted orange overall. The CAMS 58 had a lengthened cabin and could carry twelve passengers.

The small Dornier Delphin (Dolphin) first flew in 1920 and there were several versions. Some flew air services from the Swiss lakes and two Delphin IIs were operated by Bodensee Aerolloyd.

The Dornier Wal (Whale) seen here was built in 1929. There were numerous versions and many were used by Italian airlines on trans-Mediterranean services and by Luft Hansa in the Baltic. (Dornier-Werke)

Lioré et Olivier H 242

The LeO H 242 four-engined monoplane with metal hull and wooden wing was ordered as a replacement for the CAMS 53. The first example was completed in 1933 and the type went into service with Air France in 1934.

The main cabin had accommodation for ten to fifteen passengers. Most of the fourteen built remained in service until the war brought an end to Air France's Mediterranean operations.

Breguet 530 Saïgon

In 1931 Breguet obtained a licence for production of Short Calcuttas and built five. The company then produced its own designs which included two 530 Saïgon three-engined biplane flying-boats for Air Union's Marseilles-Tunis service but by the time they were ready Air France had absorbed Air Union.

The Saïgon was of all-metal construction and had three cabins. The forward cabin had eleven seats for Second-Class passengers, the midship cabin had six First-Class seats and the aft cabin had de-luxe accommodation for two or three passengers. The lavatory and baggage hold were aft of the passenger accommodation.

Latécoère 631

The Latécoère 631 was built to meet a 1936 French Air Ministry specification for a large transatlantic flying-boat able to carry 40 passengers over a stage of 6,000km (3,728 miles) against a 60km/h (37mph) headwind.

Construction was interrupted by the war but the prototype flew on 4 November 1942, was confiscated by Germany and was destroyed on Lake Constance by an Allied air attack in 1944. There appear to have been a total of eight built and Air France introduced them on Biscarosse-Fort de France services on 26 July 1947.

The Latécoère 631 was a very large six-engined monoplane with retractable floats carried by the outboard engine nacelles.

A Dornier Wal 31 forward cabin with ladder to the boarding hatch.

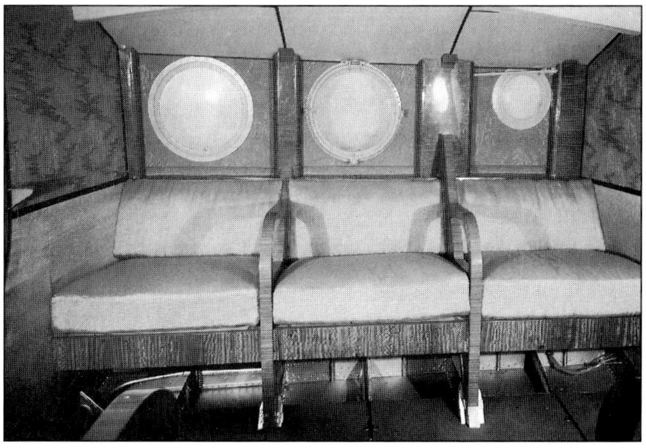

Inward-facing seats in a Dornier Wal.

A 1930 Cant 10ter, a modified version of the Cant 10ter which was designed for SISA's Trieste—Venice—Pavia—Turin service, which began on 1 April 1926 and was the first Italian air service. The bow cabin had four seats.

Normal accommodation was for 46 passengers in several cabins. There were seven two-seat cabins forward of the wing and aft of this a lavatory, baggage compartment, restaurant and bar. Beneath the wing centre section there were four four-seat cabins and aft of them two lavatories on each side and eight two-seat cabins. Behind these cabins were more lavatories (holes in the floor) and right aft a kitchen with butane-gas cooker. Passenger seats could be curtained off and converted to sleeping berths.

One of Ad Astra Aero's rather primitive Macchi-Nieuport M.9 flying-boats at Zürichhorn near Zürich. The aircraft in the background is a Macchi-Nieuport M.3 also employed by Ad Astra Aero for passenger carriage from Swiss lakes. (Swissair)

One aircraft was lost on a passenger flight on 1 August 1948 and the type was withdrawn from passenger service.

Germany
Dornier Delphin (Dolphin)

At the end of 1920 Dornier flew the first of an unusual family of small flying-boats. This was the Do Cs II Delphin, a high-wing single-engined all-metal monoplane with seats for five or six passengers.

The much improved Delphin II had accommodation for six or seven passengers and the Delphin III of 1928 had twelve or thirteen seats.

Only a small number of Delphins were produced, Ad Astra Aero (predecessor of Swissair) is believed to have used one from Swiss lakes and Bodensee Aerolloyd had two.

Passengers board a Sikorsky S-38 for the inaugural Buffalo—Toronto flight on 29 July 1929. (American Airlines)

The unusual Savoia Marchetti S.66 twin-hulled flying-boat. Each hull had seven or nine seats.

Part of one of the forward cabins of a Savoia Marchetti S.,66. It has been reported that there were four sleeping berths in the rear of each hull but no evidence that they were ever used.

View aft through a Pan American Sikorsky S-38. (Courtesy R E G Davies)

Dornier Wal (Whale)

The all-metal Dornier Wal first flew as early as November 1922 and remained in production until the mid-1930s with about 300 being built.

It was a strut-braced parasol monoplane with two engines in tandem mounted on the wing and had sponsons or sea wings which were long a trademark of Dornier.

Because of the Allied ban on German aircraft construction the early Wals were built in Italy and several Italian airlines used the type on Mediterranean services. Aero Lloyd in Germany used Wals on Danzig-Stockholm services and several operated in South America.

There were numerous versions of the Wal with different engines, wing span, weights and performance.

There was an eight-to-ten-seat cabin in the bows of the early Wals, some had inward-facing seats and the Marina-Fiat MF.5 version had accommodation for up to fourteen passengers.

Italy

Macchi-Nieuport M.18

A number of similar Macchi-

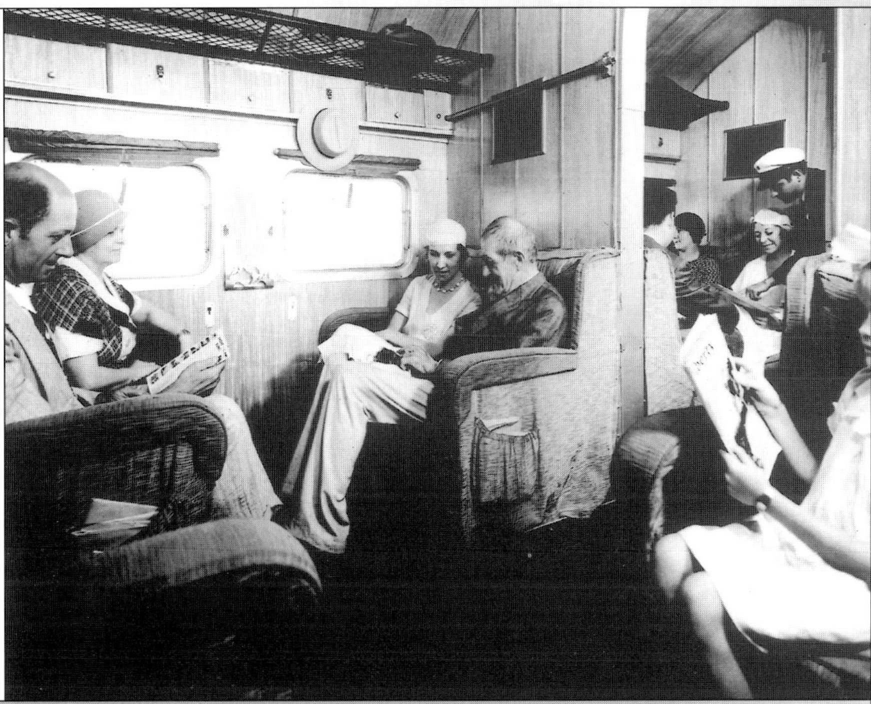

ABOVE: A NYRBA Consolidated Commodore.

RIGHT: A corner of one of the cabins of a Consolidated Commodore used by NYRBA and Pan American Airways. (Courtesy R E G Davies)

BELOW: One of Pan American Airways' large Sikorsky S-40s, the first of the Clippers.

A Sikorsky S-42 of Pan American Airways. (Pan American Airways)

Nieuport flying-boats were produced in the early 1920s. They were used by Italian airlines and by Ad Astra Aero in Switzerland.

The M.18 was typical of the series and was an unequal span biplane of wood and fabric with a single pusher engine and a small cabin for pilot and three passengers. It cruised at 90 mph.

Cant 10

In 1925 Cantieri Navale Triestino built the Cant 10 transport flying-boat and the next year it was followed by at least seven Cant 10ter which operated Adriatic services. It was of mixed construction with a single pusher engine and a bow cabin and had seats for four passengers. Some remained in service until 1937.

Meal service aboard a Pan American Airways' Sikorsky S-42. The notice above the centre window stated '4 life jackets under floor'. The cabins were separated by watertight bulkheads.

Savoia Marchetti S.55 and S.66

Savoia Marchetti built two types of most unusual configuration both of which undertook passenger service in the Mediterranean.

The first, the S.55 was built in 1924 as a torpedo bomber but adapted for passenger carriage and the second, the S.66, appeared in 1932 as a development of the S.55.

The unusual feature of these wooden monoplanes was the use of twin hulls for passenger accommodation. The S.55 had two engines in tandem above the wing and the S.66 had three engines side-by-side. The S.55 normally had accommodation for four or five passengers in each hull and the S.66 had seven or nine seats in each hull. It was reported that there were four sleeping berths in the rear of each S.66 hull but there is no evidence that they were ever used.

U.S.A.
Sikorsky S-38

Also unusual was the Sikorsky S-38 which was an eight-passenger unequal span twin-engined biplane amphibian. It had a fairly short hull and twin booms ran back from the wing to support the tail surfaces.

From 1928 NYRBA (New York, Rio & Buenos Aires Line) used some and Pan American Airways developed its Caribbean routes with a fleet of S-38s. That airline and its associates had a total of 38. The cabin had eight forward facing seats, four on each side of the central aisle. Entrance was through a hatch in the top of the hull aft of the wings.

The Martin M-130 opened trans-Pacific services for Pan American Airways in October 1936, from San Francisco to Manila.

Consolidated Commodore

The twin-engined Consolidated Commodore was chosen by NYRBA for the operation of its services between the USA and South America and the first one went into service in November 1929.

The Commodore was a strut-braced parasol monoplane with a very high standard of passenger accommodation in its spacious hull which provided cabins 8ft wide and 5ft high. Normal accommodation was for twenty-two passengers in two eight-seat compartments and two three-seat drawing rooms but 30 passengers could be carried on a Buenos Aires-Montevideo shuttle.

The NYRBA fleet was taken over by Pan American Airways when it acquired NYRBA and one Commodore was still in service, with Bahamas Airways, until 1949.

Sikorsky S-40

The first of Pan American Airways' Clippers was the Sikorsky S-40 which entered service at the end of 1931 and resembled a much enlarged S-38 with four engines but continued the twin boom tail-support structure of the earlier type.

Only three S-40s were built but they were the biggest US civil aircraft of the time and had high standard accommodation for 38 passengers, seats on the starboard side being double units. The hull was divided into several compartments and all seats may have been in facing pairs, There were window blinds and light-luggage racks.

Sikorsky S-42

The Sikorsky S-42 must be regarded as one of the great passenger carriers of the 1930s. It first flew in March 1934 and entered service with Pan American Airways on the Miami-Rio de Janeiro route in August that year.

It was a four-engined parasol monoplane which could carry its full payload of 32 passengers over a 750 miles stage. There were four cabins each with a facing pair of seats on each side. There was a galley forward and lavatory aft.

Four-abreast seating in a Pan American Airways' Martin M-130. (Courtesy R E G Davies)

The Boeing 314 was almost certainly the finest passenger flying-boat produced. The type began Pan American Airways' trans-Atlantic service on 28 June 1939 from Port Washington, New York, to Lisbon and Marseilles via the Azores.

Ten S-42 series were built, one pioneered North Atlantic survey flights and the type opened Pan Am's San Francisco-Auckland service in 1937 and in the same year the extension of the trans-Pacific route from Manila to Hong Kong.

Martin M-130

There were only three of these four-engined flying-boats but they have an important place in air transport history having operated the first trans-Pacific air services. The well-remembered China Clipper inaugurated trans-Pacific mail services on 22-29 November 1935 and Hawaiian Clipper left San Francisco on 21 October 1936 to inaugurate trans-Pacific passenger services. Maximum seating was for 41 but heavy fuel loads frequently reduced the number of available seats to about eight. High fares kept the demand low. The range of the Martin M-130 is quoted as 3,200 miles.

Boeing Model 314

Almost certainly the finest passenger flying-boat ever produced, the Boeing 314 operated the first North Atlantic passenger air services by heavier-than-air craft, from New York to Lisbon and Marseilles on 28 June 1939.

The 314 was designed to meet Pan American Airways' requirement for a transocean flying-boat and first flew on 7 June 1938. Six were ordered by Pan American followed by six Boeing 314A modified examples. Three served BOAC on wartime North Atlantic services. Their range was 3,500 miles.

The hull of the 314 had four separate compartments and maximum passenger accommodation was for 74 or 40 in sleeper configuration. The high standard of passenger accommodation is obvious in the accompanying photograph.

The wing span of the Boeing 314 was 152 feet, its maximum take-off weight 82,500lb and cruising speed 183mph.

Part of the accommodation in a Pan American Airways' Boeing 314. (Pan American Airways)

CHAPTER TWENTY HELICOPTERS

A Sabena twelve-passenger Sikorsky S-58 leaving Allee Verte Heliport in Brussels. Sabena operated eight S-58s on international services radiating from Brussels. (Sabena)

Although the helicopter is capable of providing city-centre to city-centre operations its rôle as a scheduled service vehicle has been limited. Chicago, New York, Los Angeles and San Francisco have all had local networks of helicopter services. Sabena operated international routes radiating from Brussels and Pakistan International Airlines for a time worked services in East Pakistan based on Dhaka, then Dacca.

In the United Kingdom British European Airways pioneered helicopter operations with mail and passenger services. The Penzance—Scilly service continues to be operated by British Airways.

Sikorsky S-58

This twelve-seat single-engined single-rotor helicopter first flew in March 1954 and provided higher standards of passenger comfort than previous helicopters, with lower noise and vibration levels. Sabena operated S-58s, and its Brussels—Paris service was the first international capital-to-capital operation by helicopters.

Some S-58s had separate compartments each with six facing seats and its own entrance door.

Sikorsky S-61

The S-61 has probably been the most used passenger helicopter and its two engines have removed many of the restrictions placed on single-engined types. It was the first truly practical

View forward in a New York Airways' Sikorsky S-58.

A British European Airways Sikorsky S-61N taking-off from St Just, Land's End, for Scilly. (BEA)

ABOVE: The cabin of a Sikorsky S-61N MkII. (Sikorsky courtesy R E G Davies)
BELOW: A New York Airways' fifteen-passenger single-engined twin-rotor Vertol 44B taking-off from the West 30th Street Heliport in Manhattan. (New York Airways)

helicopter in airline service and it can carry up to 30 passengers in accommodation comparable to fixed-wing types.

The S-61 was built in two versions, the S-61L landplane and the amphibian S-61N with 'boat hull' and stabilizing floats. The latter was used by Pakistan International Airlines when it opened a helicopter network in East Pakistan, now Bangladesh, in 1963, and it is used by British Airways on Penzance — Scilly services.

The cabin of the S-61 is 78in wide and 75.5in high. Cabin length is 31ft 11in.

Vertol 44B

The Vertol 44B was the first twin-rotor helicopter to be used on scheduled airline operation when New York Airways introduced it in

The 25-passenger Vertol 107. (Boeing)

May 1958. It was the first helicopter to have passenger accommodation closely resembling that of a fixed-wing airliner.

It had a cabin width of 5 ft 8in and height of 6ft 6in and the fifteen seats were in pairs on the starboard side. The aisle was along the port side which also had one backward-facing seat. The windows were curtained and the cabin was finished to a high standard.

Vertol 107

The Vertol 107, although not widely used, entered service with New York Airways in 1962 and was an important design because it had two engines and improved economics. The fuselage was amphibious and had an 8 ft 5in wide cabin accommodating up to 25 passengers in a three-abreast configuration. Like the Vertol 44B it had twin rotors and its engines were shaft-turbines.

ABOVE: The cabin of a New York Airways' Vertol 44B. This was the first type of helicopter with aeroplane-like accommodation. (New York Airways)
BELOW: Mock-up of the Vertol 107 cabin in New York Airways' colours — blue with neutral and sunburst orange. (Boeing)

Boeing Vertol Chinook

The Chinook is not known to have entered scheduled airline service but carries large numbers of passengers in the form of oil workers being taken to and from offshore platforms.

The Commercial Chinook has eleven pairs of double Economy seats separated by a 17in aisle to allow 44 passengers to be carried on 600 miles stages. There is a galley and a lavatory and the furnishing is to Boeings fixed-wing standards.

British Airways used Boeing Vertol Chinook twin-engined twin-rotor helicopters to carry off-shore workers to and from oil platforms in the North Sea. (Jim Ferguson)

Mil Mi-6P

The Soviet Union operated large numbers of helicopters, some on scheduled services but details of their operations are sparse.

At the 1967 Paris Air Show the USSR exhibited a Mil Mi-6P with 80 seats but it is not known whether it ever went into service. The Mi-6P had two 4,700hp shaft-turbines and the rotor span was 35metres (114ft 10in).

***LEFT:** The 44-seat mock-up of a Boeing Vertol Chinook. It has Boeing 727-type windows and overhead baggage bins. (Boeing)*

***BELOW:** The Mil Mi-6P with 80 seats was exhibited at the 1967 Paris Air Show. It is not known whether it ever went into passenger service. (John Stroud)*

CHAPTER TWENTY-ONE
FIRST JETLINERS AND PROPELLER-TURBINE TYPES

The world's first pure-jet airliner, the de Havilland D.H.106 Comet. (British Aerospace)

The 1950s was the period in which air transport was revolutionized by the introduction of the gas turbine into commercial aviation. The first pure-jet type was the British de Havilland Comet which operated the first jet-propelled passenger service in 1952 and the first propcllcr turbine-powered type was the Vickers-Armstrongs Viscount which entered service a year later following brief operations in 1950.

The gas-turbine added greatly to passenger comfort with its low levels of vibration and noise, by cutting flight times and enabling higher cruising levels.

Before the end of the decade the first of the 'big' jets had entered service, the Boeing 707 and Douglas DC-8 and the days of the mainline piston-engined airliners were relegated to history.

ABOVE: The cabin of a BOAC Comet 1. (de Havilland)
BELOW: A de Havilland Comet 4 of BOAC. Comet 4s operated the first North Atlantic jet services. (Hawker Siddeley)

de Havilland D.H.106 Comet

The Comet was the first jet-propelled airliner to fly and to enter passenger service. The first prototype flew on 27 July 1949 and the type entered service with BOAC on the London—Johannesburg route on 2 May 1952.

In layout the Comet was a low wing monoplane with four turbojets buried in the wing. There was accommodation for 36 passengers but the interior diameter of the cabin was slightly less than that of the Douglas DC-4.

The Vickers-Armstrongs Viscount V.630 G-AHRF about to leave Northolt on 29 July 1950 to work a BEA service to Paris. This was the first passenger service to be operated by a turbine-powered aeroplane. The Viscount had four Rolls-Royce Dart propeller-turbines.

ABOVE: Rear section of the cabin of the Viscount 630. (Vickers-Armstrongs)
BELOW: View forward through the Viscount 630. (Vickers-Armstrongs)

The Comet 1, as it became designated, brought dramatically reduced flight times and had a very low level of vibration but unfortunately three Comets broke up in flight and the type was withdrawn in 1954.

However, de Havilland and BOAC maintained their faith in the aircraft and modified Comet 4s were added to the airline's fleet in 1958 and on 4 October that year they operated the first North Atlantic jet services in each direction but with a refuelling stop on the westbound flight. The Comet 4 could accommodate 56 First-Class passengers or up to 97 in high-density layout.

Developed from the Comet 4 was the lengthened Comet 4B with clipped wings and without the pinion fuel tanks. This version could carry 92 Tourist-Class passengers and was operated by BEA and Olympic Airways.

The final development of the Comet as an airliner was the Comet 4C which combined the Comet 4 wing and Comet 4B fuselage. It could carry up to 101 passengers and entered service in 1960. Altogether, including military aircraft, there were 113 Comets.

BEA's Vickers-Armstrongs Viscount V.701 RMA Robert Falcon Scott. Viscount 701s began the first regular propeller-turbine-powered services on 18 April 1953. (BEA)

Vickers-Armstrongs Viscount

The Viscount was an outstanding airliner, not only was it the world's first propeller-turbine passenger aircraft it achieved considerable export orders and 445 were built. The first Viscount was the V.630 with 32 seats, it first flew on 16 July 1948 and on 29 July 1950 it operated the first ever propeller-turbine-powered passenger service, from London-Northolt to Paris-Le Bourget. It continued the London-Paris operation for a fortnight and then spent a similar period operating Northolt-Edinburgh services, the first domestic services anywhere with a gas-turbine-powered aeroplane.

Like all subsequent airline Viscounts the V.630 had four Rolls-Royce Dart propeller-turbines and an oval-cross section pressurized fuselage with low noise-level and almost no vibration.

Developed from the V.630 was the V.700 with increased span and 7ft 4in greater length. The production version was the V.701 and BEA ordered twenty (later 26) with five-abreast seating for 47 passengers and the type entered regular service on 18 April 1953.

ABOVE: The cabin of a BEA Viscount 701. (BEA)
BELOW: Part of the cabin of a Central African Airways' Viscount 748. The 26in by 19in windows were a major attraction. (Vickers-Armstrongs)

Pan American World Airways was the first customer for the Boeing 707. The Boeing 707 was the first of the so-called 'Big Jets'. (Boeing)

TOP: First Class lounge in a BOAC Boeing 707-436. (BOAC)

BELOW: The attractive cabin of a Boeing 707. There were passenger service units above each seat-row. (Boeing)

In the early days of Viscount operation passengers used to stand pencils on end and coins on edge on the tables and there was insufficient vibration to displace them. More appreciated by passengers were the 26in by 19in elliptical cabin windows.

Following the V.701 and other V.700 series came the V.800 with greater power and a 13 foot increase in length. It could accommodate 53 First-Class passengers, 63 Tourist-Class or 71 in high-density layout.

The Viscount V.810 series was the last development, it had uprated engines and seating for up to 75 passengers.

Boeing 707 and 720

The first of what were termed the 'big' jets was the Boeing 707 and this fine aeroplane introduced entirely new and greatly improved standards to air travel.

The 707 was a low-wing monoplane with four turbojets suspended from the wing on pylons. The internal diameter of the upper fuselage was 140.5in (11.7ft) and the first examples, the 707-120, could carry up to 179 passengers. Seating was in triple units each side of a central aisle and there was a forward lounge. There were full-length baggage racks and beneath these were passenger units with call bells, lighting and ventilation and drop-out emergency oxygen masks. Large ceiling lights could be dimmed at night and represented a deep blue sky with stars.

The second of the 'Big jets'. A Douglas DC-8 of Iberia Spanish Airlines. (Douglas)

Pan American World Airways ordered twenty Boeing 707s and began a world airline race into the jet age. The airline introduced 707s on New York — Paris services on 26 October 1958.

The original 707s were not designed to have North Atlantic non-stop range, but such range was achieved by the larger and heavier 707-300 series.

Boeing built 725 Model 707s for commercial customers including a number of specialized types, such as the 707-138 for Qantas,

B models with turbofan power had improved performance. So reliable was the Boeing 707 that some airlines scheduled daily long-haul flights while only possessing one aircraft.

There were also 154 examples of the scaled-down Boeing 720 and 720B which was a very fine aeroplane. It was supposed to be a short-haul aeroplane but the first one delivered to Pakistan International Airlines set an unbeaten world record by flying non-stop the 3,900 nautical miles from London to Karachi in 6hr 43min 51sec. Initial seating was for 165 Tourist-Class passengers.

Douglas DC-8

Slightly later than the Boeing 707 was the very similar Douglas DC-8 and again it was Pan American that started it off with an order for 25. The DC-8 was produced in several versions and the first, the DC-8-10, entered service with Delta Air Lines on 18 September 1959. Total DC-8 production was 556 and the final versions were the Super 70s with stretched fuselages, quieter engines and seats for up to 251 passengers.

Tupolev Tu-104

The second type of jetliner to enter service was the Soviet Union's Tupolev Tu-104 which began operation on the Moscow—Omsk—Irkutsk route on 15 September 1956.

The Tu-104 was based on the Tu-16 twin-turbojet bomber but with a new fuselage having an internal width of 10ft and seats for 50 passengers. There were two small cabins aft of the forward entrance and at a higher level over the spar was a buffet in which there were glass-fronted cabinets containing porcelain figures. Then came an eight-seat cabin and further aft the main 28-seat cabin.

The Tu-104 was followed by the 70-seat Tu-104A with sixteen First-Class seats forward and 54 Tourist-Class seats consisting of triple seats to starboard and double seats to port.

Interior of a Douglas DC-8. It appears to have been a Scandinavian Airlines System (SAS) aircraft.

ABOVE: The USSR's first jet airliner, the Tupolev Tu-104, went into service in 1956. (Boeing)
BELOW: The main cabin of a Tu-104. Note massive baggage racks and coat hangers. (The Aeroplane)
BOTTOM: View forward through an Aeroflot Tupolev Tu-104. The objects in the seat-backs may have been oxygen masks. (Flight)

Finally there was the Tu-104B with lengthened fuselage and 100 seats. Tu-104s remained in service until mid-November 1979.

Antonov An-10

The Antonov An-10, originally known as the Ukraina (Ukraine), was designed for high-density traffic routes and to have very good take-off and landing performance and the ability to operate from unprepared surfaces.

It was a high-wing monoplane with four propeller-turbines and three cabins seating 84 passengers three-abreast. It was first shown at Vnukovo Airport, Moscow, in July 1957 and entered service with Aeroflot in July 1959.

Ilyushin Il-18

The Il-18 was designed to meet Aeroflot's need for a high-capacity aeroplane capable of operating over a high percentage of its routes and the ability to work from modest aerodromes. It was a low-wing monoplane powered by four propeller-turbines and produced in a number of versions.

The Il-18 entered service with Aeroflot as an 80-seat aircraft with 65 seats in the main cabin, five

abreast, and a forward cabin with fifteen seats. Later versions had 73-84 and 89-111 seats. Seating was reduced in winter because of the extra space needed to stow heavy clothing. The Il-18E could carry 122 passengers in summer.

Numbers of Il-18s were exported and more than 500 were built. The first Il-18 flew in July 1957 and some are still in operation.

TOP: The propeller-turbine Antonov An-10A could operate from less developed airports.
ABOVE: The centre cabin of an 84-passenger Antonov An-10A.
BELOW: Five-abreast seating in an Aeroflot Ilyushin Il-18. (Flight)
BOTTOM: An Aeroflot Ilyushin Il-18. It had four Ivchenko AI-20 propeller-turbines and entered service in April 1959

Sud-Aviation SE.210 Caravelle

The superb Caravelle was designed to meet a specification for a medium-range transport capable of carrying a 6,000-7,000kg (13,327-15,432lb) payload for 1,600-2,000km (994-1,242miles) at 620km/h (385.2mph) with the main consideration being for operation from France to North Africa.

The first prototype flew in May 1955 and regular operation, with Air France, began in May 1959. There were numerous versions and most had accommodation for up to 80 passengers in Tourist-Class configuration.

The Caravelle was a low-wing monoplane and the

The Sud-Aviation SE.210 Caravelle was Europe's first short-range jetliner. This example was a Caravelle VIN of Indian Airlines Corporation. (Sud-Aviation)

first to have rear-fuselage-mounted engines, in this case two turbojets. It was one of the finest European airliners for many years and 282 were built.

Convair 880 and 990

The third United States builder of jet airliners was Convair, a division of General Dynamics. Their first jet type was the Convair 880, its type number derived from its speed of 880ft/min or 600mph.

ABOVE: View forward in a Finnair Caravelle 1A. (Finnair)
BELOW: The cabin of a prototype Caravelle. (Sud-Aviation)

The Convair 880 was a medium-range aeroplane designed to meet a TWA specification. In layout it resembled the Boeing 707 and Douglas DC-8, with four wing-mounted turbojets but its maximum cabin width was only 10ft 8in and it was only possible to install five-abreast seating with a total of 110 seats in all-Coach-Class configuration.

The type first flew in January 1959 and it entered service in May 1960 with Delta Air Lines on the New York—New Orleans route. The type had an impressive performance but was not a commercial success and only 65 were built.

The Convair 990 was a development of the 880 with lengthened fuselage, increased wing area and turbofan power. It was produced to meet an American Airlines specification for a transcontinental aeroplane and first flew in January 1961 but the type had many problems and extensive modification was necessary making the financial loss on the 990 one of the heaviest by a United States industrial undertaking.

The 990 could carry up to 121 Tourist-Class passengers or 158 in high-density layout. Swissair operated the type as the Coronado but the development delays cut into commercial potential and only 37 were built.

Bristol 175 Britannia

The Bristol Britannia was a superb aeroplane but it appeared too late to achieve widescale use. The original Series 100 type was intended for operation on BOAC's medium-haul routes to Africa and the East but although the first example flew in August 1952 the Britannia did not enter service until February 1957.

The Britannia had four propeller-turbines and was the first large type designed to have such power units.

The fuselage internal diameter was 139in and could accommodate up to 115 passengers in high-density configuration.

The Series 300 aircraft was designed to operate non-stop North Atlantic services and went into service with BOAC between London and New York on 19 December 1957 as the first turbine-powered type on North Atlantic services. It could accommodate 139 passengers in high-density layout. Only 89 civil and military Britannias were built.

Tupolev Tu-114

The Tupolev Tu-114 was unique, it was a very large sweptback low-wing monoplane powered by four propeller-turbines yet capable of near jet performance.

First flown in October 1957 it was then the world's largest and heaviest commercial aeroplane. It went into service with Aeroflot in April 1961.

ABOVE: The third United States producer of jetliners was Convair a division of General Dynamics Corp. Swissair operated Convair 990As under the name Coronado. (Convair)
BELOW: A corner in Taiwanese Civil Air Transport (CAT) Convair 880 Mandarin Jet. The interior was decorated in traditional Chinese style.
BOTTOM: The only other type of jetliner produced by Convair was the model 880. This is one of two Convair 880Ms that were operated by Swissair. (Convair)

The 75/115-passenger Bristol 175 Britannia was a superb aeroplane and so quiet that it was widely known as the 'Whispering Giant'. On 19 December 1957 a BOAC Britannia 312 operated the first North Atlantic service by a turbine-powered airliner, from London Heathrow to New York. (Scott Henderson Collection)

First Class seating in an El Al Britannia 313. (Bristol Aeroplane Co)

Part of the accommodation in an Aeronaves de Mexico Britannia 302. (Bristol Aeroplane Co)

A 90-passenger Tourist Class layout in a Britannia. (Bristol Aeroplane Co)

The standard aeroplane had seating for 170 passengers with 146 Tourist-Class and 24 First-Class, the latter could be replaced by sleeping berths. In international layout there where 72 Tourist-Class seats in the forward cabin and 40 First-Class in the rear cabin. Other layouts are known to have been used.

For a time Tu-114s were used on joint Aeroflot-Japan Air Lines services between Tokyo and Moscow. About 30 were built.

Vickers-Armstrongs Vanguard

The Vanguard with four propeller-turbines was a large-capacity short-haul aeroplane built for BEA and developed into a long-range type for Trans-Canada Air Lines and these two airlines were the only customers for new aircraft.

The first aircraft flew on 20 January 1959 and the type entered service with BEA in December 1960. It could carry up to 139 passengers. Interior fuselage diameter was 128in and high-density seating was six-abreast. Only 43 were sold.

Lockheed L-188 Electra

The Lockheed L-188 Electra was the only large United States passenger aeroplane to be powered by propeller-turbines, in this case four.

It was designed to meet an American Airlines specification and first flew in December 1957. It entered service with Eastern Air Lines, in January 1959.

The internal fuselage diameter of 128in made possible First-Class seating for 66, Tourist-Class for 85 and 98 in high-density layout.

The Electra suffered two structural failures and for over a year operated under severe restrictions

ABOVE: An Aeroflot Tupolev Tu-114 just after take-off from Tokyo. This very large propeller-turbine-powered aeroplane had a performance approaching that of the early jetliners. (Toshihiko Watanabe)

until modified aircraft were unrestricted in 1961. Only 170 aircraft were built which was unfortunate because it was a very nice aeroplane to fly in.

Handley Page H.P.R.3 Herald and H.P.R.7 Dart Herald

The H.P.R.3 Herald was designed as a Douglas DC-3 replacement and first flew in August 1955. It was a pressurized high-wing monoplane with retractable nosewheel-undercarriage, and four Alvis Leonides Major piston engines.

In spite of extensive sales tours it received no orders and because of the success of the Rolls-Royce Dart propeller-turbine the H.P.R.3 was lengthened and fitted with two Darts. In this form

ABOVE RIGHT: The 100/170-seat Tu-114 in intercontinental layout had 72 Tourist-Class seats in the forward cabin and 40 First-Class seats in the rear cabin. (The Aeroplane)
RIGHT: Main cabin of a Lockheed L-188 Electra. (Lockheed courtesy Derek James)

United States manufacturers only built one type of large propeller-turbine-powered airliner — the Lockheed L-188 Electra. (Scotpick)

The Vickers-Armstrongs Vanguard with four Rolls-Royce Tyne propeller-turbines could seat up to 139 passengers and was a very fine aeroplane but it was only ordered by British European Airways and Trans-Canada Air Lines. (BEA)

Looking aft through an Air Canada Vanguard. (D M Stroud collection)

the Herald was put into production as the H.P.R.7 Dart Herald.

The fuselage cross-section consisted of intersecting radii with an upper interior width of 8ft 10in and lower width of 15ft 7in. Cabin height was 6ft 4in reduced to 5ft 11in beneath the wing spar. Normal Tourist-Class capacity, four-abreast, was 50 but 56 passengers could be carried in high-density configuration.

The delay caused by using the Leonides engines and the success of the Fokker Friendship reduced the Herald's sales potential and only 50 were built.

Jersey Airlines began flying Heralds on 19 May 1961 and BEA used three on Scottish routes. The last Herald was withdrawn from service in 2000.

Fuselage mock-up of the Handley Page Herald. (Handley Page)

Avro 748 and BAe ATP (Advanced Turbo Prop)

The Avro 748, built by Hawker Siddeley and British Aerospace, was another attempt to produce a DC-3 replacement. It was put into production largely as a result of Indian requirements for a civil and military type to replace the DC-3.

The 748 was a low-wing monoplane with two Rolls-Royce Dart propeller-turbines. The fuselage was of circular cross-section with an internal diameter of 8ft. Accommodation was for 40-58 passengers four-abreast.

The second H.P.R.3 in BEA livery after conversion to H.P.R.7 Dart Herald. BEA operated three Dart Heralds on Scottish routes but not the one illustrated. (Charles E Brown)

The type first flew on 24 June 1960 and Skyways Coach-Air began 748 operations in the spring of 1962.

There were several versions including a modified military aircraft, and 381 of all models were built, 89 being assembled or built in India.

The British Aerospace ATP was to a large extent a further development of the Avro 748. It had the same overall configuration but was powered by two more powerful Pratt & Whitney Canada PW126A propeller-turbines. The ATP was slightly larger than the 748 and had accommodation for 64-72 passengers.

The ATP first flew on 6 August 1986 and 65 were built. A number flew on British Airways' Scottish routes for several years.

Fokker F.27 Friendship

Although Fokker had built a long line of airliners between the wars it was not until November 1955 that the company built another airliner of its own design. This was the Fokker F.27 Friendship designed as a DC-3 replacement.

View looking forward through the Avro 748. (A.V.Roe & Co Ltd)

One of the 748s operated by British Airways on their Highlands and Islands Division for nearly twenty years, from March 1974. (A.V.Roe & Co Ltd)

The ATP (Advanced Turbo Prop) was designed as a larger replacement for the 748, but it suffered from lack of sales and production was halted after 65 were built. (British Aerospace)

ABOVE: The cabin of a Fokker Friendship. (Fokker)
BELOW: East-West Airlines Fokker F.27 Friendship City of Tamworth. (Norman L Danvers)

The Friendship was a high-wing monoplane and its outstanding success was largely due to its two Rolls-Royce Dart propeller-turbines.

Original accommodation was for 48 passengers in four-abreast seating.

Fairchild in the United States built the type under licence and it was a Fairchild F-27 which first went into service, with West Coast Airlines in September 1958.

Numerous versions of this very successful aeroplane have been built including the Fairchild Hiller FH-227 with a stretched fuselage and seats for 56 passengers. Dutch and US production of all versions totalled 787.

CHAPTER TWENTY-TWO
THE VC10, SHORT-HAUL JETS, SUPERSONIC TRANSPORT AND WIDE-BODY AIRCRAFT

The Vickers-Armstrongs VC10 was built for BOAC. (British Aerospace)

The last three decades have brought about many marked changes in air transport. The United Kingdom built its last large subsonic airliner and with France produced the only supersonic transport to undertake prolonged service. Jet aircraft were introduced on short-stage routes and the United States brought out wide-bodied aircraft of very large capacity. Boeing absorbed McDonnell Douglas and in Europe Airbus Industrie produced a family of jet transports to rival Boeing and is working on a very large aeroplane for delivery in 2006.

View rearward in the First-Class cabin of a BOAC Standard VC10. (British Aerospace)

Vickers-Armstrongs VC10 and Super VC10

The VC10 was a superb aeroplane but by the time it was available most potential customers had equipped with Boeing 707s or Douglas DC-8s. The British type was to have better airfield performance than the United States types making it more suitable for BOAC's African and Eastern routes.

It was a low-wing monoplane with T-tail and four rear-fuselage-mounted Rolls-Royce Conway by-pass engines. Internal fuselage diameter was 138.6in and it could accommodate 139 passengers in mixed-class layout or 163 in high-density layout.

The VC10 entered service on the London—Lagos route on 29 April 1964 and quickly proved its outstanding airfield performance.

The 93-seat Tourist-Class cabin and part of the galley in a BOAC VC10. (British Aircraft Corporation)

The de Havilland/Hawker Siddeley Trident gave good service to BEA and played a major rôle in the development of Autoland. (Scott Henderson collection)

BOAC operated twelve Standard VC10s and total production was 54 including the longer-range Super VC10 which began North Atlantic operation in April 1965. The Super VC10 could carry 139 passengers in mixed-class accommodation or 163 in Economy-Class.

De Havilland/Hawker Siddeley Trident

The Trident, which went into regular operation on 1 April 1964, was designed to meet a BEA requirement for a medium-range aeroplane. Its development was prolonged partly as a result of customer changes during its design and although it did not attract widescale sales it was an important aeroplane and played a major part in the development of Autoland.

It was a low-wing monoplane with three rear-mounted engines. The Trident 1 in BEA service originally had eight First-Class seats and 76 Tourist-Class or 88 seats in an all Tourist-Class configuration. The cabin was 66ft 10in long and the internal fuselage diameter was 135.5in.

The Trident 1 had a long take-off run and a fairly high landing speed but airfield performance was improved with the Trident 2 and especially the Trident 3 which had a Rolls-Royce booster engine above the rear fuselage. Final seating was 140 in the Trident 3B.

The type served BEA and later British Airways well and the Chinese order for 33 brought the total number of Tridents built to 117.

British Aircraft Corporation One-Eleven

This was the first Western short-to-medium-range jet airliner. It first flew in August 1963 and entered service, with British United Airways, on 9 April 1965 with accommodation for 89 passengers.

The BAC One-Eleven was a low-wing monoplane with two rear-mounted turbojets. It was a very pleasant aeroplane in which to fly but caused a lot of complaints at airports by its high external noise level.

Numerous versions were built and 230 were produced in the United Kingdom before production was taken up in Romania under the title Rombac One-Eleven.

The Tourist cabin of a BEA Trident. (BEA)

ABOVE: The first of fourteen British Aircraft Corporation BAC One-Eleven short-haul jetliners for Braniff International Airways. (British Aircraft Corporation)
TOP RIGHT: A BAC One-Eleven cabin in Tourist-Class configuration. (British Aircraft Corporation)
BOTTOM RIGHT: The cabin of a Trans-Australia Airlines Boeing 727. (John Stroud)

Failure to make an early stretch of the fuselage was one reason that restricted sales.

Boeing 727

The short to medium-range Boeing 727 was regarded by Boeing as a gamble when in December 1960 it announced its intention of building the type. But it was a gamble which paid-off with more than 1.830 aircraft sold.

The layout of the 727 resembled that of the Trident, with three rear-mounted engines. Its fuselage cross-section was the same as the Boeing 707's and up to 129 passenger could be seated, but more normally the seating was for less than 100.

During its production the 727 was constantly improved and the 727-200, which entered service

A Northwest Airlines Boeing 727. More than 1,800 Boeing 727s were built. (Boeing)

A Douglas DC-9-10 of Swissair.

ABOVE: Tourist-Class cabin of a Douglas DC-9-10.
BELOW: One-Class cabin of a Boeing 737-400. (Boeing)

with Northwest Orient Airlines in December 1967 had a 20ft increase in fuselage length.

The Boeing 727 was a truly great aeroplane with a fine airfield performance and many have been re-engined with turbofans to improve performance and meet the noise abatement levels now introduced.

Douglas DC-9-10

The 80-90-seat Douglas DC-9-10 medium-range twin-jet first flew in February 1965 and entered service with Delta Air Lines that December. It is a low-wing monoplane with T-tail and two rear-fuselage-mounted turbofans. This model of which 137 were built led to a long line of development into many different models achieving large sales.

Boeing 737

The Boeing 737 is one of the most successful airliners ever built. It first flew in April 1967 and began service in February 1968. It has the same

One of British Airtours' Boeing 737-200s. (Boeing)

The Aérospatiale-British Aircraft Corporation Concorde is the only supersonic transport to have been in sustained service. It went into service in January 1976 and one made a New York — London flight in less than three hours. This British Airways' example had Singapore Airlines livery on the port side. (Scott Henderson collection)

fuselage cross-section as the 707 and 727 and two wing-mounted turbofans and a low-mounted tailplane.

The 737-200 has single-class seating for 115 passengers and a maximum of 130.

Subsequent models are the 737-300, -400, -500, -600, -700, -800 and -900 and the type is still in production after more than 30 years. Accommodation in the latest model, the 737-900, has increased to 177 in two classes. Well over 3,000 Boeing 737s have now been sold.

Aérospatiale-British Aircraft Corporation Concorde

The Anglo-French Concorde supersonic transport is one of only two types of supersonic airliners built and the only one to continue in sustained service. It has been in operation since January 1976 and is operated on North Atlantic routes by Air France and British Airways.

This technically brilliant design has accommodation for 100 passengers, four-abreast, in two cabins and cruises at Mach 2 with a top speed of 1,400mph and a maximum cruise level of 60,000ft.

The slender fuselage is mounted on a delta wing of complex geometry and there are four Rolls-Royce-Snecma Olympus turbojets in pairs beneath the wing.

By mid-May 1967 airlines had taken options on 74 aircraft but eventually only Air France and British Airways took delivery and there were only sixteen production aircraft.

Boeing 747

In April 1966 Boeing announced that Pan American World Airways had placed an order for 25 Boeing 747s. This is a vast aeroplane with a wing span a third greater than that of the Boeing 707 and it is 50% longer than the earlier type. Its original weight of 683,000lb was more than double that of the 707 which it resembles, but with a forward upper deck. The basic layout was for 363 passengers with an envisaged maximum of about 490, but the 747SR for Japan Air Lines has 525 seats. When the 747 came into service in January 1970 its originally quoted weight had already risen to 712,000lb and later models were to exceed 875,000lb.

A British Airways' Concorde with original furnishing. (British Airways)

A Boeing 747-136 of BOAC. (Boeing)

Nine-abreast seating in a Swissair Boeing 747-200B. The cabin width is just over 20ft. (Courtesy Swissair)

View forward through the Economy-Class cabin of a Swissair Boeing 747-200B with a film being shown. (Courtesy Swissair)

The 747's main cabins are over 20ft wide. Interior layouts vary but British Airways' 747-400s have five rows of First-Class seats in the nose with five seats each side plus two pairs at the rear of the cabin, then comes the Club-World cabin with 23 seats, and aft of the lavatories and stairway to the upper deck are three cabins with a total of 328 World-Traveller (Economy-Class) seats mostly with triple seats on each side and four-abreast seats between the two aisles. The main deck can have additional World Traveller seating and

The front (First-Class) cabin of a Boeing 747 with spiral stairway to upper deck. (Boeing)

Business-Class cabin of a Boeing 747-400 with straight stairway to upper deck. (Boeing)

A Swissair Douglas DC-10-30. (Douglas)

there are 32 Club World seats on the upper deck. British Airways has more than 50 of this version in service.

Several versions of the 747 are in service including the 747SP (Special Performance) with a shorter fuselage and a very long range. In 1976 it set a non-stop record for commercial aircraft by flying 16,560km (8,940miles) from Seattle to Cape Town. The latest and the longest range 747-400, with winglets, can carry up to 568 passengers in two-class configuration and has a design range of 13,491km (7,284miles) with 420 passengers in three classes.

More than 1,290 Boeing 747s have been sold.

ABOVE: Looking aft through a Douglas DC-10-30. (George Gerster courtesy Swissair)
BELOW: Cabin of an Airbus A300. (Sud-Aviation)

Douglas DC-10-30

The Douglas DC-10 was the first three-engined wide-body airliner. It was designed to meet an American Airlines' specification for a 250-passenger aeroplane capable of carrying its full load between Chicago and the US West Coast. American Airlines introduced the 208 seat DC-10-10 domestic model in August 1971. The same month, United Air Lines

The Airbus A300 was the first type of twin-engined wide-body aircraft. (Airbus)

A Court Line Lockheed L-1011 Lockheed TriStar. (Lockheed)

ABOVE: First-Class lounge and bar in a Gulf Air TriStar. (Courtesy Gulf Air)

BELOW: Eight-abreast Economy-Class seating in a Gulf Air TriStar. (Courtesy Gulf Air)

began DC-10 service between Los Angeles and Washington DC.

The long-range overwater DC-10-30 flew in June 1972 and entered service that December with Swissair on North Atlantic routes.

The DC-10 is a swept wing aircraft with two wing-mounted engines and a third at the base of the fin and rudder. Coach-Class seating is nine-abreast.

Including military KC-10s, a total of 446 DC-10s was built.

Lockheed L-1011 TriStar

Like the Douglas DC-10, the Lockheed TriStar was designed to meet American Airlines' requirement for a 250-passenger aeroplane capable of carrying its full load between Chicago and the US West Coast.

The TriStar resembled the DC-10 in layout, size and weight but the third engine was positioned in the rear fuselage with an S intake duct forward of the fin.

In July 1981 the TriStar was certificated to land with zero ceiling and 200m horizontal visibility and was the only jet transport cleared for Cat IIIA automatic landing.

Probably the most luxurious interior in an airliner was that in Gulf Air's 221-seat TriStars. They had extra large seats, thick carpets, a large lounge, a shopping boutique and a library. In First-Class there

A Swissair Douglas DC-9-81. (Douglas)

were gold leather seats, large tables and a stand-up bar. A total of 250 TriStars were produced.

Airbus A300

Airbus Industrie was formed in the 1960s as a consortium to develop a twin-engined wide-body airliner to fill a gap in the market. The original members were Aérospatiale, Deutsche Airbus, Hawker Siddeley (now BAE Systems), and CASA of Spain with Fokker and Belairbus (Belgium) as associates.

The first product was the large A300 which was the first twin-engined wide-body airliner. It first flew as a 250-passenger aeroplane in October 1972 and the modified A300B2 entered service with Air France in May 1974. Production ended in May 1984 after 249 had been built and Airbus began production of the improved 267-344-seat A300-600 with 109ft 9in long cabin. It accommodates 26 First- and 241 Tourist-Class passengers in a mixed class configuration, the latter eight-abreast. In all-Tourist configuration, there are up to 285 seats eight abreast and the charter layout can seat 344, nine-abreast.

The aircraft entered service in April 1996 with Swissair. The A300-600 is still in production with 271 having been ordered by 1 January 2000.

The A300 series are very fine aircraft operationally and have a high standard of passenger comfort and the success of the first Airbus has led to an entire family of airliners, making Airbus presently the only major rival to Boeing.

At the end of March 2000, a total of 3,667 aircraft of all Airbus types had been ordered, with 2,150 in operation.

Douglas DC-9-81

After Douglas Aircraft merged with McDonnell Aircraft in 1967, the new McDonnell Douglas company continued to develop the highly successful DC-9 twin-jet, but its stretched DC-9-81 was re-designated MD-81. The MD-81 with lengthened fuselage and many other improvements over earlier models first flew in October 1979 with accommodation for a maximum of 172 passengers in single-class configuration compared with the

Economy-Class layout in a Douglas DC-9-80. (Douglas)

An Egyptair Boeing 767.(Boeing)

LEFT: Looking aft in a Boeing 767. (Boeing)
BELOW LEFT: Looking aft in a Boeing 757 Economy-Class cabin. (Boeing)

DC-9-30's 115. In mixed-class layout the MD-81 could accommodate 137 instead of the DC-9-30's 93.

The MD-81 entered service, with Swissair, in October 1980. The MD-81 was followed by a number of new variants culminating in the MD-95 a 100-passenger regional jetliner which is now being marketed as the Boeing 717 following the Boeing take-over.

Boeing 767 and 757

In 1978 Boeing announced that it was proposing to develop a new family of large transport aircraft. These were the Models 757 and 767 with the latter flying first. The 767 is a high-capacity wide-bodied aeroplane and the 757 a short-to medium-range type intended to replace the Model 727.

The two types have almost identical flight decks and very similar handling qualities. Both types are twin-engined with the engines in pods beneath the wing.

Maximum seating in the 767-300 version is 350 eight-abreast, while the 757-300 can seat up to 289 in high-density configuration.

A later development of the 767 is the 767-400ER (extended range) which was due to enter service in mid-2000. This version has accommodation for up to 245 passengers in three classes and has a design range of 5,580 nautical miles.

Airbus A310

The second Airbus type was the medium-to

British Airways' Boeing 757 Dover Castle. (Boeing)

long-range A310 which closely resembles the A300 but is thirteen frames shorter and has seats for up to 280 passengers but more normally 200-234.

Of great advantage on the European sectors is the aircraft's ability to climb straight to 41,000ft keeping it above most other traffic and shortening flight times.

At the end of March 2000, a total of 260 A310s had been ordered and 248 were in service.

ABOVE: The overhead bins in a Boeing 757. (Boeing)
BELOW: The cabin of a Swissair Airbus A310-221.

British Aerospace 146

Unlike the aircraft already described in this chapter the BAC 146 is a high-wing monoplane. It is a regional jet with four underwing turbofans, creating low noise which allows unrestricted night time operation. The first example flew in September 1981, was designated BAe 146-100 and had accommodation for 70-94 passengers. This was followed by the 85-112-seat 146-200

The A310 was the second type of Airbus. (Airbus Industrie)

A British Aerospace 146 operated by AirUK now KLM uk.

Cabin of a BAe 146-300 (British Aerospace)

and the 100-128-seat 146-300.

These have now been redesignated Avro RJ 70, 85 and 100 based on their typical seating. Over 200 have been built, but production has now ceased.

Avions de Transport Regional ATR42 and ATR 72

These Franco-Italian twin-engined high-wing monoplanes are propeller-turbine-powered regional airliners with accommodation for 42 and 72 passengers respectively.

Although typical seating is 42 for the ATR42 it can carry up to 50 passengers four-abreast. The ATR42 first flew in August 1984 and the ATR72 in October 1988.

Orders for both types total more than 500, many for United States airlines.

Fokker 100

The Fokker 100 is a 100-seat regional twin-jet developed from the F.28 Fellowship of May 1967. It is a low-wing monoplane with T-tail and rear-mounted turbofans.

Before the collapse of Fokker in 1997 a total of 283 had been built as well as 48 Fokker 70s, a shortened development of the Fokker 100, for 70-79 passengers.

The cabin of an ATR42. (D M Stroud collection)

The longer cabin of the ATR72. (D M Stroud collection)

An American Eagle Franco-Italian ATR42. (Avions de Transport Regional)

McDonnell Douglas MD-11

The MD-11 wide-body trijet was developed from the DC-10 with a lengthened fuselage and accommodation for up to 405 passengers. Externally the MD-11 can be recognized by its winglets and revised tail cone. The MD-11 first flew in January 1990 and 200 were built.

An example of the MD-11's interior is that adopted by Alitalia. It has twelve First-Class (Top-Class) seats in pairs forward, 35 Business-Class aft of the Top-Class and two cabins with a total of 156 Economy-Class seats up to nine-abreast. Its Business Class cabin includes a stand-up bar.

Airbus A320 Family

In April 1988 Air France introduced the twin-engined single-aisle Airbus A320 and it was at the time the most technically advanced subsonic airliner. Typical two-class accommodation is for

ABOVE: All-Economy-class cabin of an AirUK Fokker 100. Carpets and seats were deep blue. (Reeve Photography courtesy AirUK)
BELOW: A Fokker 100 of AirUK.

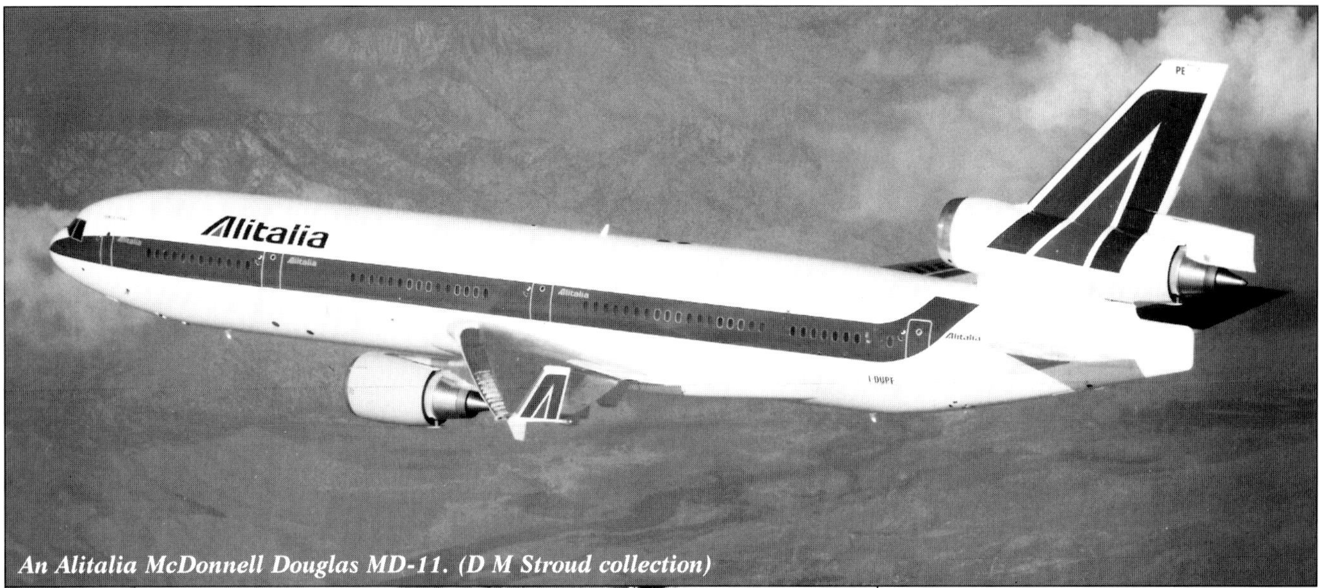

An Alitalia McDonnell Douglas MD-11. (D M Stroud collection)

ABOVE: First-Class cabin of an MD-11. No seat-belts are visable so this may be a photograph of a mock-up. (D M Stroud collection)

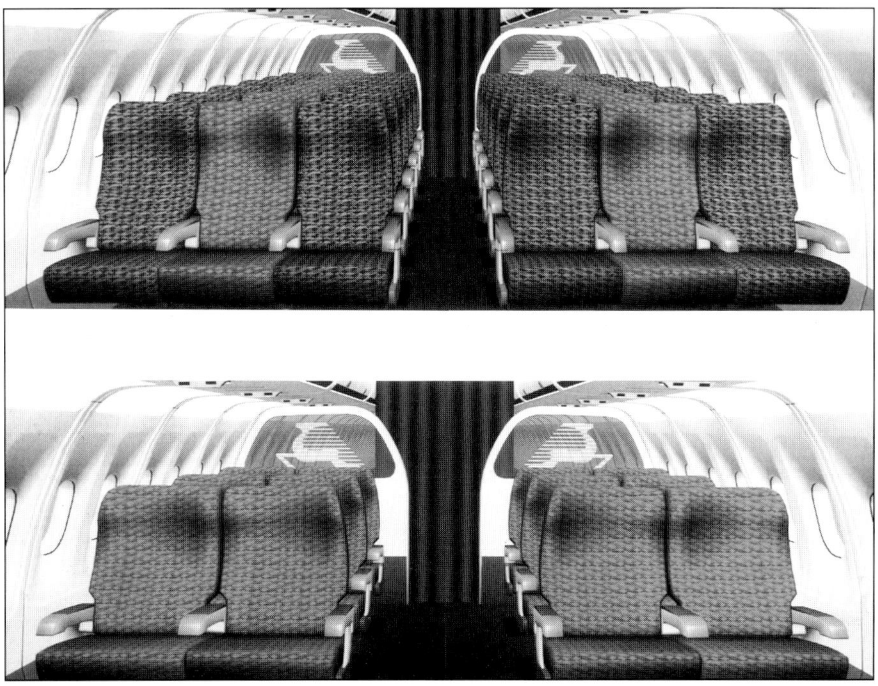

Economy(top) and First-Class (lower) seating in an Airbus A320. These are computer-aided designs. (Airbus)

150 seats but maximum high-density seating is for 179.

Airbus describes the A320 as having 'fly-through-computer controls' and the aircraft was the first to have its entire flight envelope safeguarded. It is protected against stalling, excesssive speed, airframe overloading and windshear. The pilots' controls are sidesticks in place of the hitherto used yokes. Twenty-one A320-100s were built and followed by the A320-200 of higher weight and fitted with winglets.

In 1993 Airbus flew the lengthened A321 with 186-220 seats. This was followed by the shorter A319 with 124-142 seats, becoming the shortest member of the family until the even shorter A318 with 107-117 seats is introduced into service in 2003.

At the end of March 2000, 120 orders had been received for A318s, 574 for A319s, 1,321 for A320s and 315 for A321s.

Airbus A330

A completely new large capacity medium-to long-range twin-engined Airbus entered service at the end of 1993, this was the A330-300 with seating for 295-440. It was followed into service in April 1998 by the shorter fuselage 256-293 seat A330-200.

An Airbus A319 of British Airways. (Airbus)

The maximum internal fuselage width is 208.2in. First-Class seating is in pairs outboard of each aisle with a third pair between the aisles. Business-class is the same but with triple centre seats. Standard Economy-Class seating is eight-abreast and a high-density layout has nine-abreast seating. Overhead luggage bins are fitted each side and above the centre seats.

Range with 335 passengers is 4,750 nautical miles and the maximum take-off weight is 467,400lb. Two hundred and sixty-four A330s had been ordered by the beginning of 2000.

RIGHT: Business-Class seats in a SriLankan Airlines Airbus A330-200. (Airbus)

BELOW: Business-Class seats in a SriLanken Airlines Airbus A330-200, with large seatback displays and video-players in the armrests. (Airbus)

Airbus A340

The A340 shares many components with the A330 but is powered by four engines, is heavier and has greater range. There are two versions, the A340-200 with seating for 203-262 and slightly longer A340-300 which has the same maximum weight of 558,900lb, accommodation for 295-335 passengers and a range of 6,750

A Cathay Pacific Airways' Airbus A330. (Airbus)

BELOW: First-Class sleeper seats in a Kuwait Airways Airbus A340. (Airbus via D M Stroud)

BELOW: Part of the Upper-Class cabin of a Virgin Atlantic Airways' Airbus A340. (Virgin Atlantic Airways)

BELOW: First-Class seats with personal television in a Gulf Air Airbus A340. (Airbus)

BELOW: First-Class cabin of an Airbus A340. (Airbus via D M Stroud)

An Airbus A340 of Virgin Atlantic Airways. (Virgin Atlantic Airways)

nautical miles with 295 passengers. Stretched versions with accommodation for up to 372 passengers are the A340-500 and -600 which are due to enter service in 2002.

At the end of March 2000 291 A340s had been ordered and 174 were in service.

First-Class cabin of a Boeing 777. (Boeing)

Boeing 777

Boeing's long-range high-capacity Model 777 in its -300 version can for the present be claimed as the longest airliner ever built with a length of 242ft 4in which can cause space problems at certain airports.

The Boeing 777 can accommodate up to 440 in the -200 version and up to 550 passengers in an all-Economy high-density version of the 777-300.

A British Airways Boeing 777. (Scott Henderson)

Computer-generated image of the Airbus A380 in Air France livery. (Airbus)

The new type is twin-engined and externally resembles earlier Boeings but a new feature is wing folding to reduce its 199ft 11in span to 155ft 3in. It is said to incorporate more advanced technology than any other Boeing. Several hundred have been ordered. The 777-200 first flew in June 1994 and the -300 in October 1997.

Airbus A380

In December 2000 Airbus after receiving orders for 50 aeroplanes, announced that it would put the A380 into production. Before that announcement the type was designated A3XX.

The A380 will be the largest passenger aeroplane ever built, with a span of 79.8m (261ft 10in), a length of 73m (239ft 6in) and a maximum take-off weight of 548,000kg (1,208,000lb) or an optional 560,000kg (1,235,000lb).

The A380 will be a low-wing monoplane powered by four engines of 68,000 or 70,000lb thrust and the heavier cargo version will have 75,000lb thrust engines.

Passenger accommodation in the A380-100 will be in two full-length decks with seating for 555 passengers in a three-class layout or 854 all-Economy-Class. The projected A380-200 will have up to 960 seats and the A380-300 is expected to seat 1,000. Fuselage diameter will be 6.95m (22ft 10in).

Stairways will connect the two decks and possible features include shops, showers and exercise rooms. Singapore Airlines is to be the first operator and is scheduled to receive its first A380 in February 2006.

BELOW, LEFT and RIGHT: Two views of the interior mock-up of the Airbus A380. (Airbus)

CHAPTER TWENTY-THREE USSR AND CIS

The first modern Soviet airliner was the ten-passenger ANT-35 (PS-35) of 1935. Only a few were built but some operated Moscow — Riga — Stockholm services.

In the early years of air transport in the Soviet Union many of the aircraft used were imported and large numbers of single-engined biplanes produced in the USSR were in service.

RIGHT: The cabin of a Czechoslovak Il-14. The seats could be reclined and there were pull-out tables.

BELOW: The Ilyushin Il-12 entered service with Aeroflot in 1948 and was intended to replace the Li-2s (Soviet licence-built Douglas DC-3s). A development of the Il-12 was the very similar Il-14 which was built in the USSR, Czechoslovakia and East Germany. (John Stroud)

The Ilyushin Il-62 of 1963. (Scott Henderson)

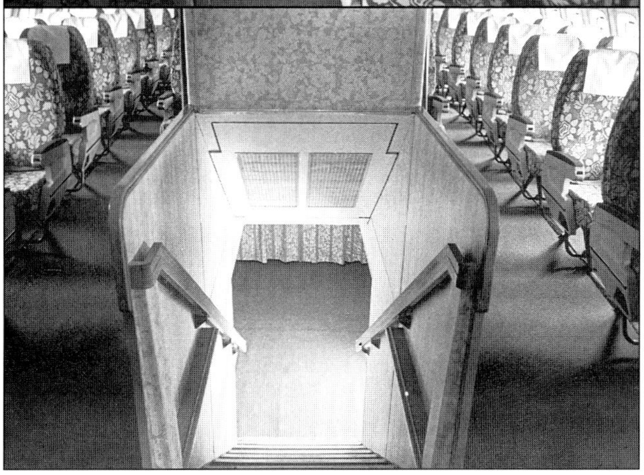

TOP LEFT: Six-abreast seating in an Aeroflot Il-62. (Flight)
LEFT UPPER AND CENTRE: Looking aft through an Aeroflot Il-86.(Boris Korzin courtesy D M Stroud)
BOTTOM LEFT: Stairway from main deck to lower deck baggage compartment of an Il-86. (Boris Korzin courtesy D M Stroud)

Numbers of A N Tupolev ANT-9 three-engined monoplanes were used including those of Deruluft from 1932 on the Königsbierg — Moscow route. The ANT-9 looked like a high-wing Junkers-Ju52/3m or a Fokker trimotor with corrugated metal skin. In 1932 the ANT-9-M-17 was produced with only two engines.

The first Soviet 'modern' airliner was the twin-engined ANT-35 (PS-35) which flew at the end of 1935 and was exhibited at the 1936 Paris Air show.

The USSR built 6,157 Douglas DC-3s under licence and for a number of years these together with C-47s supplied under Lease Lend maintained the main services of Aeroflot. The post-Second World War replacements were the Ilyushin Il-12 and Il-14. In the mid-1950s the USSR embarked on a major modernization of its air transport fleets. Some of the types produced are described in Chapter Twenty-one.

ANT-35 (PS-35)

The ANT-35 was a twin-engined low-wing monoplane with retractable main undercarriage but non-retractable tailwheel. It had accommodation for ten passengers and was to some extent the equivalent of the Lockheed Electra. It is thought that only a small number was produced but the type is known to have operated Moscow — Riga — Stockholm services from July 1937. Strangely for a Soviet airliner it appears not to have had any airframe de-icing

Passengers boarding an Aeroflot Ilyushin Il-86 by the front of the three airstairs. (Boris Korzin courtesy D M Stroud)

Ilyushin Il-12

The Ilyushin Il-12 was designed as a replacement for the Li-2 which was a Soviet-built Douglas Dakota. The Il-2 first flew in 1946 and was a twin-engined low-wing monoplane with retractable nose-wheel undercarriage and seating for probably, 27 passengers in nine rows of three and later on international services twenty-one seats. Large scale use of Il-12s probably began in May 1948. Large numbers were built and some exported to Czechoslovakia and

The 32-seat aft cabin of a CSA Tupolev Tu-124V.

A Tupolev Tu-124V of Ceskoslovenské Aerolinie (CSA). (Sania Foto)

An Aeroflot Tupolev Tu-134

Poland. Some Il-12s seem to have been converted on the production line to the inproved Il-14.

Ilyushin Il-14

The Il-14 was essentially similar to the Il-12 but initially had 24 seats in six rows of four abreast. In 1958 the seating was increased and Il-14s were in service with 28, 32 and 36 seats. The passenger cabin was 8.9m (29ft 2.25n) long, 2.67m (8ft 9in) wide and 1.94m (6ft 4.5in) high.

Very large numbers of IL-14s were built in several versions and production also took place in Czechoslovakia (as the Avia-14) and in the German Democratic Republic as the VEB Il-14P. The Avia-14 Super had a pressurized cabin and seats for up to 42 passengers.

Ilyushin Il-62

The Ilyushin Il-62 was the first Soviet four-jet airliner to go into production. It is a low-wing monoplane with T-tail and engines in pairs on either side of the rear fuselage. First flying in January 1963, it was designed to replace Tu-114s on intercontinental routes.

Cabin width is 3.45m (11ft 3.25in) and five seating configurations are known. Standard Economy-Class seating is for 186 all six-abreast, Tourist-Class seating is for 168 and First-Class 114.

Emergency oxygen is not carried for each occupant but one pilot has to remain on oxygen for all time spent above 4,000m (13,123ft). There are a number of versions including the 198-seat Il-62M-200 with increased range of 8,000km (4,970miles) with a maximum payload of 23,000kg (50,706lb) and full fuel reserves.

Interior of an Aeroflot Tu-134 with television sets. (V Sozinov courtesy D M Stroud)

Ilyushin Il-86

The Ilyushin Il-86 which entered service with Aeroflot in December 1980 is the first Soviet wide-body airliner and it has maximum seating for 350 passengers nine-abreast. The cabin is 5.7m (18ft 8.5in) wide and 2.61m (8ft 7in) high.

It is a very big aeroplane with a span of 48m (157ft 8in) and a length of 59.54m (195ft 14in). It has four 13,000kg (20,246lb) thrust engines in pods beneath the wing and is certificated for CAT III automatic landing.

The Tupolev Tu-144 was the first supersonic transport to fly, on 31 December 1968, but it did not go into service until November 1977 and was soon withdrawn. (Air Portraits)

There are two main decks. The upper deck is divided into three cabins mostly with nine-abreast seating. Passengers board by a set of airstairs to each cabin via the lower deck where they can deposit their baggage. The lower deck also houses the galley which has a food lift between the forward and midship cabins There are six lavatories at the rear.

About 100 Il-86 aircraft have been built, and there is an Il-96 development with triplex fly-by-wire controls.

Tupolev Tu-124 and Tu-134

Aeroflot had a requirement for a jet airliner with about half the capacity of the Tu-104B with an improved airfield performance. Tupolev produced the new type as the twin-engined Tu-124 which resembled a scaled-down Tu-104. It first flew in June 1960.

The original accommodation was for 44 and the subsequent Tu-124V version was equipped for 56 passengers.

The exterior fuselage width was 2.9m (8ft 10.5in) and the cabin height 1.9m (6ft 2in). In both versions there were double seats on each side of the centre aisle, being divided into front, midship and rear cabins.

There are thought to have been about 100 Tu-124s produced and there was passenger criticism of the high level of noise in the midship and aft cabins. Production was phased out in favor of the Tu-134 which was essentially an improved Tu-124 with increased capacity, T-tail and two rear-mounted engines. Standard accommodation is provided for 72 Tourist-Class passengers in two cabins.

Tupolev Tu-144

The Tu-144 was the Soviet supersonic airliner, actually flying before Concorde. It was a delta-winged aeroplane with four engines in pairs beneath the wing. It was designed to accommodate 140 passengers in two cabins with 99 seats in the aft cabin with 45 triple seats and twelve double seats on the starboard side and all double seats on

View looking aft through a Tupolev Tu-144. (V Sozinov courtesy D M Stroud)

A Malév Hungarian Airlines Tupolev Tu-154 B2. (Scott Henderson)

the port side. The front cabin had six rows of seats, again triple seats to starboard and double to port. There were six lavatories and two galleys. The forward cabin could have eighteen First-Class seats.

The Tu-144 first flew on 31 December 1968, Mach 2 was achieved in June 1969, and the type entered Aeroflot service on 1 November 1977. It was withdrawn in May 1978 after the loss of one aircraft and retired from service.

Tupolev Tu-154

The Tupolev Tu-154 was designed to replace An-10s, Il-18s

ABOVE LEFT: Six-abreast seating in a 164-seat Aeroflot Tu-154. (V Sozinov courtesy D M Stroud)
LEFT: The cabin of a LOT Antonov An-24V. (LOT)
BELOW: A Polskie Linie Lotnicze - LOT Antonov An-24V. This was the Soviet equivalent of the Fokker Friendship and was built in large numbers.

A line-up of Aeroflot Yakovlev Yak-40s. (TASS)

and Tu-104s. It first flew in October 1968 and is a low-wing monoplane with T-tail and three rear-mounted engines. The passenger accommodation was designed for 128-167. Scheduled service began in August 1972.

About 900 Tu-154s have been built and the improved Tu-154M is still in production. The aircraft is in widescale use.

An interior view looking aft of an Aerocarribean Yakovlev 40. (Scott Henderson)

Antonov An-24

The An-24 was another DC-3 replacement. Aeroflot required a 32-40-seat aircraft for medium-stage routes, which would provide reliable, safe and economic service in a wide range of climates. During development the seating was increased to 44 and later 48.

A Yakovlev Yak-42D of Cubana Cuban Airlines. (Scott Henderson)

A line up of Antonov An-2s. Thousands of these utility aircraft were used by Aeroflot. (Scott Henderson)

In appearance, size and weight the An-24 closely resembles the Fokker F.27 Friendship, being a high-wing monoplane with retractable nosewheel undercarriage and two propeller-turbines.

The An-24 first flew in December 1959 and it is believed that more than 500 are still in service.

Developments of the An-24 are the An-26, 30 and 32, and the type is built in China as the Xian Y7.

Yakovlev Yak-40 and 42

The Yak-40 first flew in October 1966 and was the world's first jet transport designed to operate from short unprepared runways. It is a low-wing monoplane with three rear-mounted engines and accommodation for 24-31 passengers. One version has single-seats to port and double-seats to starboard and strangely baggage racks only along the starboard side. It is believed that about 1,000 have been built and many remain in service with numerous airlines in the CIS and elsewhere.

The Yak-42 is of similar layout but can carry up to 120 passengers. It was designed as an An-24 and Tu-134 replacement.

Antonov An-2 and Ilyushin Il-100

In spite of being a single-engined biplane and more than 50 years old it seems that many thousands of some 18,000 built are still working local services in Russia and the Commonwealth of Independent States.

The An-2 has exceptional airfield performance and is operated on wheel, ski and float undercarriages. It can carry up to twelve passengers in a rather austere cabin.

In February 2000 it was announced that Ilyushin was completing design of a replacement for the An-2. This is to be the twelve-seat Il-100, a high-wing monoplane with two propeller-turbines. It is being designed for a life of 60,000hr and to cruise at about 280kt It will require less than 400m for take-off. It is aimed to fly the Il-100 two years after the go-ahead which has not yet been given.

An interior view looking forward in an Aeroflot An-2. (Scott Henderson)

BIBLIOGRAPHY

Aeroflot: An Airline and its Aircraft, R E G Davies and Mike Machat

Aeroplani S.I.A.I. 1915-1935, Giorgio Bignozzi and Roberto Gentilli

Airlines of the United States since 1914, R E G Davies

Airspeed Aircraft since 1931, H A Taylor

Armstrong Whitworth Aircraft since 1913, Oliver Tapper

Avro Aircraft since 1908, A J Jackson

Boeing Aircraft since 1916, Peter M Bowers

Bristol Aircraft since 1910, C H Barnes

Curtiss Aircraft 1907-1947, Peter M Bowers

de Havilland Aircraft since 1909, A J Jackson

European Transport Aircraft since 1910, John Stroud

Fokker Aircraft Builders to the World, Thijs Postma

Fokker Verkeersvliegtuigen, René de Leeuw editor

General Dynamics Aircraft and their Predecessors, John Wegg

Handley Page Aircraft since 1907, C H Barnes

Images of America. Sikorsky, John W R Taylor

The International Directory of Civil Aircraft 1999/2000, Gerard Frawley

Jetliners in Service since 1952, John Stroud

Lockheed Aircraft since 1913, René J Francillon

Lufthansa: An Airline and its Aircraft, R E G Davies and Mike Machat

LZ 127 Graf Zeppelin, Dr Ing W Von Langsdorff

McDonnell Douglas Aircraft since 1920, (2 vol), René J Francillon

Pan American: An Airline and its Aircraft, R E G Davies and Mike Machat

Nostalgia Pan Americana, Don Thomas

Saab Aircraft since 1937, Hans G Andersson

Shorts Aircraft since 1900, C H Barnes

Silent Swift Superb: The Story of the Vickers VC10, Scott Henderson

Soviet Transport Aircraft since 1945, John Stroud

The Ford Story A Pictorial History of the Ford Tri-Motor 1927-1957, William T Larkins

The World's Airliners, Peter W Brooks

Tupolev Aircraft since 1922, Bill Gunston

Vickers Aircraft since 1908, C F Andrews and E B Morgan

Zeppelin: The Story of a Great Achievement, Harry Vissering

Zeppelin Rigid Airships 1893-1940, Peter W Brooks

INDEX OF AIRCRAFT

SCOVAL PUBLISHING LTD